Lecture Notes in Mathematics

Volume 2385

This series reports on new developments in all areas of mathematics and their applications - quickly, informally and at a high level. Mathematical texts analysing new developments in modelling and numerical simulation are welcome. The type of material considered for publication includes:

1. Research monographs
2. Lectures on a new field or presentations of a new angle in a classical field
3. Summer schools and intensive courses on topics of current research.

Texts which are out of print but still in demand may also be considered if they fall within these categories. The timeliness of a manuscript is sometimes more important than its form, which may be preliminary or tentative. Please visit the LNM Editorial Policy (https://drive.google.com/file/d/1MOg4TbwOSokRnFJ3ZR3ciEeK s9hOnNX_/view?usp=sharing)

Titles from this series are indexed by Scopus, Web of Science, Mathematical Reviews, and zbMATH.

Steve Awodey

Cartesian Cubical Model Categories

 Springer

Steve Awodey (ORCID)
Pittsburgh, PA, USA

ISSN 0075-8434   ISSN 1617-9692   (electronic)
Lecture Notes in Mathematics
ISBN 978-3-032-08729-4   ISBN 978-3-032-08730-0   (eBook)
https://doi.org/10.1007/978-3-032-08730-0

Mathematics Subject Classification: 55U35, 18N45, 18F20, 03B38, 55P05

This work was supported by and Air Force Office of Scientific Research (FA9550-21-1-0009, FA9550-20-1-0305, FA9550-15-1-0053).

This Springer imprint is published by the registered company Springer Nature Switzerland AG
The registered company address is: Gewerbestrasse 11, 6330 Cham, Switzerland

If disposing of this product, please recycle the paper.

In memoriam F. William Lawvere

Preface

These notes were begun in 2014 during a Thematic Trimester at the Institut Henri Poincaré in Paris, which provided plenty of stimulation in a most congenial environment. They were finally completed in 2022 during the COVID-19 pandemic lockdown, which provided plenty of time with few distractions. I am particularly grateful to the several research institutes that provided generous support and many useful interactions in the intervening years of composition: the Stockholm Logic Group, the Newton Institute at Cambridge, the Hausdorff Institute in Bonn, and the Oslo Centre for Advanced Study. The Department of Philosophy at Carnegie Mellon University has also supported this project, if perhaps somewhat ambivalently, and for that I am grateful univalently.

Pittsburgh, USA   Steve Awodey
August 2025

Acknowledgments Foremost, I am indebted to Thierry Coquand for sharing his ideas in conversations at the IHES in Paris, online during the COVID-19 pandemic, at the CAS in Oslo, and on several other occasions going back to the IAS in Princeton. In many of the same places, André Joyal has provided patient advice and illuminating lectures. The writings of Mike Shulman and Emily Riehl have also been extremely helpful, as have many discussions with both, and some late comments from the latter. I have learned much from my past and present CMU colleagues Mathieu Anel, Reid Barton, Marc Bezem, Ulrik Buchholtz, Jonas Frey, Bas Spitters, Andrew Swan, and the late Pieter Hofstra; I am especially grateful for their patience through many revisions and delays. Many other people have given good advice over the long period of these investigations, including Bjorn Dundas, Peter Dybjer, Marcelo Fiore, Richard Garner, Nicola Gambino, Dan Licata, Peter LeFanu Lumsdaine, Per Martin-Löf, Ieke Moerdijk, Andy Pitts, Christian Sattler, Thomas Streicher, Benno van den Berg, and undoubtedly others that I am forgetting. An anonymous referee also provided many useful comments and suggestions, for which I am very grateful. I am also grateful to the computational higher type theory group around Robert Harper at CMU, including his students Carlo Angiuli, Evan Cavallo, Favonia, and Jon Sterling, for sharing their insights and challenging me to clarify my thoughts. For research stays during which some of this work was conducted, I thank in particular the Oslo Centre for Advanced Studies and the Institut des Hautes Études Scientifiques. Finally, I must acknowledge my debt to the late Vladimir Voevodsky, whose profound contributions advanced the subject far beyond my original expectations.

This material is based upon work supported by the Air Force Office of Scientific Research under awards number FA9550-21-1-0009, FA9550-20-1-0305, and FA9550-15-1-0053. I am extremely grateful to program officer Tristan Nguyen for the valuable support that he has provided to this research effort over many years.

Competing Interests The author has no competing interests to declare that are relevant to the content of this manuscript.

Contents

Chapter 1
Introduction

Recent years have seen renewed interest in the cubical approach to abstract homotopy theory. This contrasts with the more familiar and widespread simplicial approach, using which many sophisticated and powerful tools have been developed, such as simplicial model categories [31], quasi-categories [45], and higher toposes [56]. Of course, some early work like the original papers of D. Kan, [48, 49], employed cubical sets, and some researchers such as [21] and [43] have developed such methods further in a more modern style, but they are swimming against the tide.

The current interest in the cubical approach arises from connections with the formal system of *type theory* for the purpose of computerized proof checking, cf. [11]. Unlike previous cubical models of homotopy theory, however, the cubes being used for this purpose are generally assumed to be closed under finite products; we call such cube categories *Cartesian*. This is a natural enough assumption to make for cubes, but one that has somehow escaped serious consideration—but for two notable exceptions: in A. Grothendieck's famous letter to D. Quillen, and the accompanying 600 page manuscript *Pursuing Stacks* [39], such cubical sets make an appearance as *test categories*, which model the homotopy category of spaces in a particular way. In fact, the Cartesian cubes studied here are *strict* test categories in the terminology of op.cit., meaning that the geometric realization functor preserves finite products [24]. The more familiar category of "monoidal" cubical sets used since Kan is also a strict test category provided one includes *connections* [57], but this is not necessarily Cartesian. The second source for Cartesian cubical sets is F.W. Lawvere, who proposed them as a model for homotopy theory in lectures, and in public and private correspondence, but never (to my knowledge) published anything on the subject. Among their advantages, he stressed the *tinyness* of the 1-cube, or "interval" I, which indeed plays a role in the current theory—although perhaps not exactly the one envisioned by him. The reader familiar with Synthetic Differential Geometry [51] will perhaps recognize the analogy between I and that theory's tiny "object of infinitesimals" D, for which the tangent bundle of a smooth

S. Awodey, *Cartesian Cubical Model Categories*, Lecture Notes
in Mathematics 2385, https://doi.org/10.1007/978-3-032-08730-0_1

space M is given synthetically by the exponential M^D, just as the path space of a space X in the present theory is the exponential X^I. Under this analogy, the current development aptly becomes "synthetic homotopy theory".

We can define *the* Cartesian cube category $\Box$ to be the Lawvere algebraic theory of bipointed objects, the opposite of which is therefore the category of finite, strictly bipointed sets $\mathbb{B} = \Box^{\mathrm{op}}$. Thus $\Box$ is the free finite product category with a bipointed object $[0] \rightrightarrows [1]$. Our homotopy theory will be based on the category of *Cartesian cubical sets*, which is the category of presheaves on $\Box$,

$$\mathsf{cSet} = \mathsf{Set}^{\Box^{\mathrm{op}}}$$

and thus consists of all *covariant* functors $\mathbb{B} \to \mathsf{Set}$. Among these, there is an evident distinguished one, namely that which "forgets the points", and it is represented by the generating 1-cube $[1]$,

$$I = \Box(-, [1]) : \mathbb{B} \longrightarrow \mathsf{Set} .$$

In cubical sets, the bipointed object $1 \rightrightarrows I$ turns out to have the (non-algebraic) property that its two points have a trivial intersection.

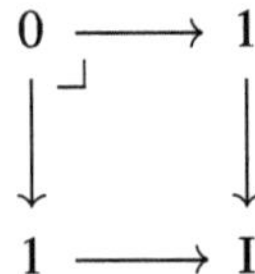

We call such an object in a topos an *interval*, and in a sense to be made precise, this is the universal one. Other categories of Cartesian cubical sets have a canonical comparison to this one, relating their respective homotopy theories.

For the purpose of homotopy theory, namely, this interval provides a good cylinder $X + X \rightarrowtail I \times X$ for every object X, as well as a good pathobject $X^I \twoheadrightarrow X \times X$ for every *fibrant* object X. The notion of fibrancy here is determined by the interval I in terms of *paths* $I \to X$, and is a generalization of the *path-lifting* condition from classical homotopy theory, suitably modified for this setting. We formulate it using the now-standard notion of a Quillen model structure:

Definition 1.1 ([62]) A *Quillen model structure* on a (bicomplete) category $\mathcal{E}$ consists of three classes of maps $C, \mathcal{W}, \mathcal{F}$ satisfying the conditions:

1. $(C, \mathcal{W} \cap \mathcal{F})$ and $(C \cap \mathcal{W}, \mathcal{F})$ are weak factorization systems,
2. $\mathcal{W}$ has the 3-for-2 property: if any two sides of the triangle $e = f \circ g$ are in $\mathcal{W}$, so is the third.

For the interval $1 \rightrightarrows I$, we have the mono $\partial : 1 + 1 \rightarrowtail I$ as one of two basic cofibrations C giving rise to all the others, in a certain sense. The other basic one is the diagonal $\delta : I \rightarrowtail I \times I$, which is a special cofibration that, together with ∂, determines both $\mathcal{F}$ and $\mathcal{W}$ just from the conditions (1) and (2) in the definition (which we have restated in a form due to [46]). Condition (1) has recently been termed a *premodel structure* by Barton [15], and its verification in our setting is fairly routine, occupying less than the first half of the paper. Condition (2) is where all the work is, and where our treatment is most likely to be of interest to the expert. We shall summarize those aspects below, but let us say now that the model structure is not the one determined by the method of [26], nor is it Reedy in the sense of [63], although the Cartesian cube category $\square$ is "generalized Reedy" in the sense of [17].

Having identified $\square$ as a strict test category, why not simply use standard tools to determine the test model structure on cSet, making it equivalent to the standard homotopy theory of spaces? Because *we are mainly interested in how the model structure relates to the interpretation of type theory*. Specifically, we wish to investigate the relationship between the ingredients of a Quillen model structure and certain standard constructions in type theory, in order to better understand the somewhat mysterious connection between the two.

The first models of homotopy type theory used the standard Kan-Quillen model structure on simplicial sets, cf. [12, 50]. Much subsequent work has also relied on classical methods, including M. Shulman's *tour de force* result that every Grothendieck ∞-topos admits a model of HoTT with a univalent universe [68]. This means that all of the results in the *Homotopy Type Theory* book [70] hold, not only in the standard model in "spaces," i.e. simplicial sets, but also in any such higher topos. In particular, the univalence *axiom* of V. Voevodsky is actually *true* in all such models. There is, however, a mismatch between such models of the univalence axiom and the design and implementation of computer systems based on type theory. Taken as an axiom, univalence blocks the normalization algorithm which forms the basis of type theoretic computation. Voevodsky recognized this, and conjectured (roughly) that the system with the univalence axiom admitted an interpretation into the system without it, in a way that would restore effective computation.

A version of this "homotopy canonicity conjecture" was finally verified a decade later by T. Coquand and collaborators [19, 28]. One key insight that apparently led to their success was the "change of shape" from simplicial to cubical sets.[1] Some aspects of Coquand's work were undoubtedly informed by homotopy theory, but much of it was driven by type-theoretic considerations: normalization, canonicity, constructivity, etc. Subsequent work on computational systems of univalent type theory (such as [5, 53, 60]) also used intuitions from basic homotopy theory (and some of the terminology), but without even attempting to verify the model category axioms. Of course, this research had a very different aim, namely the provision of a constructive system of type theory with univalence, which would facilitate its

---

[1] As suggested by Bezem and Coquand [18]. Whether this alone is essential is still a matter of debate; arguably, it was rather the algebraic aspect underlying the "uniform Kan filling" condition that made the break-through possible. Whether the cubical shape is essential to *that* will perhaps be determined by recent work on an algebraic simplicial approach by Gambino and Henry [33] and van den Berg and Faber [71].

implementation in a computer proof system. Once that was accomplished, there was no need to determine whether a Quillen model structure was also lurking in the background; it simply remained a mystery that the ingredients required for a computational system of univalent type theory seemed to align with the basic concepts of abstract homotopy theory.

It was C. Sattler who first recognized that a computational implementation of univalent type theory such as [28] contained everything required to determine a Quillen model structure, cf. [66]. An earlier result in this direction had been given by Gambino and Garner [32], who showed that the basic system of type theory with identity types not only interpreted into a weak factorization system (as had been shown by Awodey and Warren [12]), but that it actually *required* such a structure for its sound interpretation—essentially by constructing a weak factorization system from the system of type theory itself (P. LeFanu Lumsdaine subsequently used higher inductive types to construct a second weak factorization system within homotopy type theory in [55], making another step toward a full model structure). The relationship between the full system of univalent type theory and a full Quillen model structure is somewhat more subtle—and part of the present investigation— but the mystery of *why* the tools of model category theory seemed to work so well for constructing systems of univalent type theory is at least partially resolved by the insight that the type theory is apparently describing the same kind of structure as do certain model categories; namely, that of a *higher topos*. So while there was no reason to expect *a priori* that the work on computer proof systems would have any relevance to homotopy theory, the methods developed for those purposes have now acquired such relevance nonetheless.

These new methods include various species of cubical sets with different combinatorial and homotopical properties (see [24] for a survey), some still unknown, as well as various composition, filling, and uniformity conditions with as yet unclear relationships to homotopical algebra, cf. [5, 19, 25, 28, 60]. It is worth noting, for those not familiar with both, that translating between the language of type theory and that of model categories is by no means routine, nor is the converse anything like typesetting a commutative diagram in LaTeX. (Indeed, the limits of such translation are a matter of current investigation, with the question of how to handle in type theory the coherences arising in higher category theory at the very forefront of current research.)

The particular category of Cartesian cubes considered here has been studied by the author in lectures, notes, and papers since 2013, with various different box filling conditions (e.g. [6]). The condition explored in the present work, which we call *unbiased partial box filling*, was apparently first considered by Coquand [29], but was later abandoned in favor of a monoidal one in [19], and then modified to one depending on the presence of connections in [28]. The unbiased approach was resurrected and studied intensely in type theory by R. Harper and his students in [2–4, 23], culminating in [5]. These type theoretic constructions are analyzed here in terms of model categories for the first time, doing for the system of Cartesian cubical type theory roughly what [35, 66] did for the system in [28] (although, in this case, the model category was developed in parallel with the type theory, rather than after the fact).

Specifically, we ultimately show that the category of Cartesian cubical sets admits a Quillen model structure $(C, \mathcal{W}, \mathcal{F})$ with unbiased fibrations as the class $\mathcal{F}$ and the cofibrations C axiomatized to allow for variations, including additional structure on the basic cube category $\square$, and adjustments in the filling conditions. For the purely Cartesian case, this model structure has been shown *not* to be Quillen equivalent to spaces (see [13]), and so it is not the test model structure. It is nonetheless significant as the model structure resulting from the natural interpretation of type theory. Moreover, since our proofs are given in elementary diagrammatic form, the results will also hold in other categories of Cartesian cubical sets, including those with connections, reversals, etc. Indeed, part of our motivation was to apply the results obtained here, *mutatis mutandis*, in two other settings: realizability, and equivariant filling. The former (underway in [1]) imposes a condition of constructivity, about which we will say a bit more shortly. The latter (underway in [13]) is based on an unpublished result due to Sattler showing that an additional equivariance condition on the unbiased fibrations suffices to turn this model structure into the test model structure.

The possibility of an entirely constructive verification of the Quillen model category axioms is a consequence of the aforementioned constructive interpretation of univalent type theory labored over by Coquand and his collaborators, and it has applications for the homotopy theory of presheaves and sheaves that stand to be explored further (but cf. [30]). The important uniformity condition on the Kan filling operations is closely related to E. Riehl's *algebraic model structures*, cf. [64], and gives rise to a notion of *structured fibration* that admits classification, in the sense of classifying spaces, by means of what we here call *classifying types*. These classifying types, which are derived both from type theory and the *notion of fibered structure* introduced in [68], are used to construct universal objects of various kinds: families, cofibrations, (trivial) fibrations, and ultimately a universal fibration $\dot{\mathcal{U}} \twoheadrightarrow \mathcal{U}$, which acts like an object classifier in higher topos theory, but with a stricter universal property. Our work shows that having such classifying types can be useful, e.g., when "changing the base" from one slice category $\mathcal{E}/X$ to another $\mathcal{E}/Y$ along a map $f : Y \rightarrow X$, or along a more general geometric morphism $f^* \dashv f_* : \mathcal{F} \rightarrow \mathcal{E}$.

Another application of the constructivity of the model structure is the computation of homotopy invariants from a constructive proof. This was merely a theoretical possibility until quite recently, when a breakthrough by Ljungström [54] finally allowed the computer system Cubical Agda of [73] (which is based on the results just mentioned of Coquand et al.) to compute the value of a closed term $\beta : \mathbb{Z}$ from a proof in homotopy type theory that $\pi_4(S^3) \cong \mathbb{Z}/\beta\mathbb{Z}$, which had been done by hand 10 years earlier at the IAS by Brunerie [22]. Realizability models of type theory based on constructively proven model structures should also have applications in computational homotopy theory.

One way to verify that our model structure is entirely constructive would be to formalize the proofs below in a proof assistant such as Agda. While this could be of interest for the practice of translating model category proofs into type theory, in principle one would learn very little that is not already known, since the model structure given here already underlies a computational interpretation of type theory that has been fully formalized and verified (namely, that in [5]). Although our definitions and proofs do not parallel those in ibid. in the way that a proper formalization would, the associated interpretation of type theory will be plainly visible to the experts.

Let us now make this more explicit, as we outline the contents of the paper (references to the literature occur at the corresponding points of the main text). After defining the Cartesian cubical sets and establishing some basic facts about them in Chap. 1, Chap. 2 specifies the *cofibrations* axiomatically, as a class of monomorphisms classified by a universal one $t : 1 \rightarrowtail \Phi$. This permits using the associated polynomial endofunctor $P_t : \mathsf{cSet} \to \mathsf{cSet}$ (which is forced to be a monad by the axioms for cofibrations), to give an algebraic weak factorization system with the cofibrations as the left maps and the (retracts of) P_t-algebras as the maps on the right, which we define to be the *trivial fibrations*. Since the monad is fibered, the factorization system is stable under change of base, which we use to derive the familiar diagonal filling characterization of the trivial fibrations in algebraic form, and relate this to the uniform filling condition from type theory. The polynomial monad $P_t : \mathsf{cSet} \to \mathsf{cSet}$ is related to the type theoretic partiality- or lifting-monad, and generalizes the partial map classifier from the early days of topos theory. Indeed, since the results of this chapter generalize to an arbitrary elementary topos $\mathcal{E}$, we can develop them in that setting.

In Chap. 3, the *fibrations* are defined in terms of the trivial fibrations via the Joyal-Tierney calculus of pushout-products and pullback-homs. A "biased" version using the two endpoints $\delta_0, \delta_1 : 1 \rightrightarrows \mathrm{I}$ is given first, before specifying the "unbiased" version in terms of the generic point $\delta : 1 \to \mathrm{I}^*\mathrm{I}$ in the slice category $\mathsf{cSet}_{/\mathrm{I}}$, namely the diagonal $\mathrm{I} \to \mathrm{I} \times \mathrm{I}$. Specifically, a map $f : X \to Y$ in cSet is defined to be an unbiased fibration if its pullback to $\mathsf{cSet}_{/\mathrm{I}}$ has the right lifting property against all maps of the form $c \otimes_I \delta$ where $c : C \rightarrowtail Z$ is a cofibration over I and the pushout-product with δ is formed in $\mathsf{cSet}_{/\mathrm{I}}$.

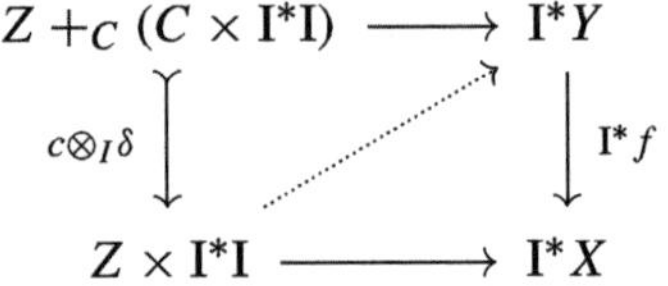

The two weak factorization systems of cofibrations and trivial fibrations, and trivial cofibrations and fibrations, are assembled formally into a Barton premodel structure in Chap. 4, where the *weak equivalences* are determined and related to *weak homotopy equivalences*: maps that induce isomorphisms in the homotopy category by precomposition. The 3-for-2 axiom is then reduced to a technical condition dubbed the *fibration extension property*, the proof of which is deferred. This concludes Part 1, and attention shifts to establishing the fibration extension property.

Part 2, consisting of Chaps. 5–8, is essentially a 60 page proof of a lemma. It seems entirely likely that a more direct proof could be given, dispatching the entire second part of the notes. Even in that event, however, the work done in Part 2 would remain worthwhile, for this is where an implicit construction of a model of (homotopy) type theory occurs: The Frobenius property in Chap. 5 establishes the interpretation of Π-types of fibrations along fibrations, and thus the right properness of the model structure, by an entirely new diagrammatic argument derived from one originally given in type theory. In Chap. 6 we construct the classifying types for fibration structure and use them to give a new construction of a universal fibration $\dot{\mathcal{U}} \twoheadrightarrow \mathcal{U}$. This is where the tinyness of the interval I plays an unexpected role, and a related axiom on the cofibrations is discovered. Chapters 7 and 8 make implicit use of the model of type theory emerging in the background, and contribute new diagrammatic proofs of two fundamental facts about it: in Chap. 7 an equivalence extension property is established which is closely related to the univalence of the universal fibration $\dot{\mathcal{U}} \twoheadrightarrow \mathcal{U}$, and in Chap. 8 that property is used to finally establish the fibration extension property, which is seen to be equivalent to the statement that the base object $\mathcal{U}$ is fibrant. In sum, then, the missing 3-for-2 property of the premodel structure from Part 1 is proven in Part 2 by constructing a fibrant, univalent universe of fibrant objects.

The main novelties in the construction are, in summary: (i) the polynomial algebraic weak factorization system of cofibrations, determined by a cofibration classifier; (ii) the notion of an *unbiased* fibration, defined using a generic point in the slice category over the interval; (iii) a new construction of the Hofmann-Streicher universe in presheaves; (iv) the use of *classifying types* of fibration structures, along with the tinyness of the interval, to construct the universe if fibrations; (v) a new proof of the realignment property for fibrations, using the condition that the pathspace functor preserves cofibrations; and finally, (vi) the axiomatic form of the argument on the basis of the structure $(\Phi, \mathrm{I}, \mathcal{V})$, which was first formulated in [9]. To be sure, some of these aspects have been used, in syntactic form, in the construction of models of type theory, but no other Quillen model category construction uses them all in this way. We also note that the purely diagrammatic proofs given for the Frobenius condition, the equivalence extension property, and the fibration extension property are entirely new (as is the proof in the appendix that cSet classifies intervals).

One thing that we learn from the exercise is that one can get quite far in constructing a model of type theory in a *premodel* category, without assuming a fibrant universe, its univalence, or even the presence of a universe at all! Conversely, our results suggest that the presence of a fibrant, univalent universe in such homotopical semantics in a premodel structure is not just necessary for a full model of univalent type theory, but actually suffices for a full Quillen model structure. In this sense, a model of HoTT is equivalent to a Quillen model category of a certain kind—namely, one that presents a higher topos.

Chapter 2
Cartesian Cubical Sets

There are many different categories of cubes $\square$ that can be taken as a site for homotopy theory, cf. [24, 37], and indeed several different ones have recently been explored in connection with cubical systems of (homotopy) type theory, including [5, 19, 25, 28, 60], to name only a few. The model structure developed here is intended to work with any of these, insofar as they are *Cartesian*, in the sense that the indexing cubes $[n] \in \square$ are closed under finite products $[m] \times [n] = [m + n]$. Rather than working axiomatically, though, we shall work in the initial such category, which we call *the Cartesian cube category* $\square$, defined as the free finite product category on a bipointed object $\delta_0, \delta_1 : 1 \rightrightarrows \mathrm{I}$.

2.1 Cartesian Cubes

Definition 2.1 The objects $[n]$ of the *Cartesian cube category* $\square$, called n-cubes, are finite sets of the form

$$[n] = \{0, x_1, \ldots, x_n, 1\},$$

where the $x_1, \ldots, x_n$, are arbitrary but distinct elements, and $0, 1$ are further distinct, distinguished elements. The arrows,

$$f : [m] \to [n],$$

are arbitrary bipointed maps $f' : [n] \to [m]$ (note the variance!). Thus $\mathbb{B} = \square^{\mathrm{op}}$ is the category of finite, strictly bipointed sets.

As a Lawvere theory, the arrows $f : [m] \to [n]$ in $\square$ may also be regarded as n-tuples of elements from the set $\{0, x_1, \ldots, x_m, 1\}$. These can be generated under

S. Awodey, *Cartesian Cubical Model Categories*, Lecture Notes
in Mathematics 2385, https://doi.org/10.1007/978-3-032-08730-0_2

composition from faces, degeneracies, permutations, and diagonals (see [61] for further details).

Definition 2.2 The category cSet of *Cartesian cubical sets* is the category of presheaves on the Cartesian cube category $\square$,

$$\mathsf{cSet} \;=\; \mathsf{Set}^{\square^{\mathrm{op}}}.$$

It is of course generated by the representable presheaves $\mathsf{y}[n]$, to be written

$$\mathsf{I}^n = \mathsf{y}[n]$$

and called the *geometric n-cubes*.

Note that the representables I^n are also closed under finite products, $\mathsf{I}^m \times \mathsf{I}^n = \mathsf{I}^{m+n}$. We write I for I^1 and 1 for I^0, which is terminal. We will need the following basic fact about the cubes I^n in cSet.

Proposition 2.3 (Lawvere)   *The n-cubes* I^n *are* tiny, *in the sense that the endofunctor* $X \mapsto X^{\mathsf{I}^n}$ *is a left adjoint.*

(See [52] on such "amazing right adjoints".)

Proof It clearly suffices to prove the claim for $n = 1$. For any cubical set X, the exponential X^{I} is a "shift by one dimension",

$$X^{\mathsf{I}}(n) \cong \mathrm{Hom}(\mathsf{I}^n, X^{\mathsf{I}}) \;\cong\; \mathrm{Hom}(\mathsf{I}^{n+1}, X) \cong X(n+1). \qquad (2.1.1)$$

Thus X^{I} is given by precomposition with the "successor" functor $\square \to \square$ with $[n] \mapsto [n] \times [1] = [n+1]$. Precomposition always has a right adjoint, which in this case we shall write as:

$$(-)^{\mathsf{I}} \;\dashv\; (-)_{\mathsf{I}}\,.$$

We call X_{I} the I*th-root of* X. $\qquad\qquad\qquad\qquad\qquad\qquad\qquad\qquad\quad\square$

The following is used to calculate the root X_{I}. A similar fact holds for the generic object in the object classifying topos $\mathsf{Set}[X] = \mathsf{Set}^{\mathsf{Fin}}$ and related categories used in the theory of abstract higher-order syntax [58].

Lemma 2.4 *For the representable functor* $\mathsf{I} = \mathsf{y}[1]$ *in* cSet, *we have* $\mathsf{I}^{\mathsf{I}} \cong \mathsf{I} + 1$.

Proof For any $[n] \in \square$ we have:

$$(\mathsf{I}^{\mathsf{I}})(n) \cong \mathsf{I}(n+1) \cong \mathrm{Hom}(\mathsf{I}^{(n+1)}, \mathsf{I}) \cong \square([n+1],[1]) \cong \mathbb{B}([1],[n+1]) \cong n+3.$$

On the other hand,

$$(I + 1)(n) \cong I(n) + 1(n) \cong \mathrm{Hom}(I^n, I) + 1 \cong \mathbb{B}([1], [n]) + 1 \cong (n + 2) + 1.$$

The isomorphism is natural in n. $\qquad\square$

Corollary 2.5 *For any cubical set X, the Ith-root X_I may be calculated as follows.*

$$X_I(n) \cong \mathrm{Hom}(I^n, X_I)$$

$$\cong \mathrm{Hom}((I^n)^I, X)$$

$$\cong \mathrm{Hom}((I^I)^n, X)$$

$$\cong \mathrm{Hom}((I + 1)^n, X)$$

$$\cong \mathrm{Hom}(I^n + \tbinom{n}{n-1}I^{n-1} + \cdots + \tbinom{n}{1}I + 1, X)$$

$$\cong X_n \times X_{n-1}^{\binom{n}{n-1}} \times \cdots \times X_1^{\binom{n}{1}} \times X_0 .$$

The exponential X^I will be called the *pathobject* of X and plays a special role. As seen in (2.1.1), it classifies "paths" in X: the "points" or 0-cubes $p \in (X^I)_0$ in the pathobject correspond uniquely to 1-cubes $p \in X_1$, the "endpoints" of which $p_0, p_1 \in X_0$ are given by composing with the "evaluation" maps

$$\epsilon_0, \epsilon_1 : X^I \rightrightarrows X$$

at the points $\delta_0, \delta_1 : 1 \rightrightarrows I$, where $\epsilon = X^\delta$. Higher cubes $c : I^{n+1} \to X$ are similarly paths between lower cubes $c_0, c_1 : I^n \to X^I \rightrightarrows X$. Note that, as a left adjoint, the pathobject functor $X \mapsto X^I$ also preserves all *colimits*.

We shall need the following two facts concerning the interaction of cubes I^n, pathobjects X^I, and the base change functors associated to a map $f : X \to Y$ in cSet, namely,

$$f_! \dashv f^* \dashv f_* : \mathsf{cSet}/_X \longrightarrow \mathsf{cSet}/_Y .$$

Lemma 2.6 *The pushforward functor along any map $f : X \to Y$ preserves pathobjects; for any object $A \to X$ over X, the pathobject of the pushforward f_*A is (canonically isomorphic over Y to) the pushforward of the pathobject,*

$$(f_*A)^I \cong f_*(A^I) .$$

Proof Over X, the pathobject A^I of $A \to X$ is A^{X^*I}, where I is the constant family $X^*I = X \times I \to X$. The claim is true for any constant family X^*C, with C in place of I, as the reader can easily verify using the Beck-Chevalley condition. $\qquad\square$

Lemma 2.7 *The pulled-back interval $1^*I = 1 \times I \to I$ in $\mathsf{cSet}/_I$ is also tiny.*

Proof Since the interval $I = y[1]$ is representable, the slice category $\mathsf{cSet}/_I$ is also a category of presheaves, namely over the sliced cube category $\square/_{[1]}$,

$$\mathsf{cSet}/_I \;=\; \mathsf{Set}^{\square^{op}}/_{y[1]} \;\cong\; \mathsf{Set}^{(\square/_{[1]})^{op}}\,.$$

However, since $\square$ does not have all finite limits, the sliced index category does not have all finite products, and so we cannot simply repeat the proof from Proposition 2.3. But as in that proof, we do have a "successor" functor

$$s_{[1]} : \square/_{[1]} \to \square/_{[1]}\,,$$

resulting from the "predecessor" natural transformation $s \Rightarrow 1_\square$ given by the projection $I \times X \to X$. Evaluating s at each object $f : [n] \to [1]$ in $\square/_{[1]}$, we obtain a commutative diagram:

$$
\begin{array}{ccccc}
s[n] & \xrightarrow{\ \cong\ } & [1]\times[n] & \xrightarrow{\ p_n\ } & [n] \\[2pt]
{\scriptstyle sf}\big\downarrow & & & & \big\downarrow{\scriptstyle f} \\[2pt]
s[1] & \xrightarrow[\ \cong\ ]{} & [1]\times[1] & \xrightarrow[\ p_1\ ]{} & [1]
\end{array}
\qquad (2.1.2)
$$

We can then set $s_{[1]}(f) = p_1 \circ sf = f \circ p_n$. As in the foregoing proof, we can then calculate the values of the adjoints on presheaves, associated to $s_{[1]}$,

$$s_{[1]!} \dashv s_{[1]}{}^* : \widehat{\square/_{[1]}} \longrightarrow \widehat{\square/_{[1]}}$$

to be, successively,

$$s_{[1]!}(X) = I^*I \times X\,,$$

$$s_{[1]}{}^*(X) = X^{I^*I}\,.$$

The first equation follows from the observation that the diagram (2.1.2) is a pullback, and so the object $s_{[1]}(f) : s[n] \to [1]$ of $\widehat{\square/_{[1]}}$ given by the evident composite is just $I^*I \times f$, and the diagram itself represents the counit map $(I^*I \times f) \to f$ over I. The second line then follows by adjointness, as does the fact that we have a further right adjoint, namely, the I^*I*th-root*:

$$s_{[1]*}(X) =: X_{I^*I}\,.$$

$$\square$$

Chapter 3
The Cofibration Weak Factorization System

To build a model structure on the presheaf category of cubical sets, one can simply take as the cofibrations *all* of the monomorphisms in cSet, but for some purposes, it is convenient to know what is actually required of the cofibrations (see Appendix A for an example). As nothing in this chapter depends on the specifics of cubical sets, the results are formulated for an arbitrary elementary topos $\mathcal{E}$. To begin, the following axioms are assumed for cofibrations.

Definition 3.1 (Cartesian Cofibrations)   A class C of *Cartesian cofibrations* in a topos $\mathcal{E}$ is a class of monomorphisms satisfying the following conditions:

(C0)   The unique map $0 \to X$ is always a cofibration.
(C1)   All isomorphisms are cofibrations.
(C2)   The composite of two cofibrations is a cofibration.
(C3)   Any pullback of a cofibration is a cofibration.

We also require the cofibrations to be classified by a subobject $\Phi \hookrightarrow \Omega$ of the standard subobject classifier $\top : 1 \to \Omega$ of $\mathcal{E}$:

(C4)   There is a terminal object $t : 1 \to \Phi$ in the category of cofibrations and cartesian squares.

Four further axioms for Cartesian cofibrations will be added later as they are needed: (C5) and (C6) early in Sect. 4.1, (C7) later in Sect. 4.4, and a final one (C8) in Sect. 7.4. The axioms are collected (and renumbered slightly) in Appendix A.

Cofibrations will be written

$$c : A \rightarrowtail B .$$

S. Awodey, *Cartesian Cubical Model Categories*, Lecture Notes
in Mathematics 2385, https://doi.org/10.1007/978-3-032-08730-0_3

3.1　The Cofibrant Partial Map Classifier

Consider the polynomial endofunctor $P_t : \mathcal{E} \to \mathcal{E}$ determined by the cofibration classifier $t : 1 \rightarrowtail \Phi$ (see [34]). We will write the value of this functor at an object X as

$$X^+ := \Phi_! \, t_*(X) = \sum_{\varphi : \Phi} X^{[\varphi]} . \tag{3.1.1}$$

The reader familiar with type theory will recognize the similarity to the "partiality" or "lifting" monad of [59]. When all monos are cofibrations, so that $\Phi = \Omega$, the object X^+ agrees with the partial map classifier $\widetilde{X}$ from topos theory, cf. [44]. We may therefore regard X^+ as the object of *cofibrant partial elements* of X, as we now explain.

Since $t : 1 \rightarrowtail \Phi$ is monic, $t^* t_* \cong 1$, so X^+ fits into the pullback square

$$
\begin{array}{ccc}
X & \rightarrowtail & X^+ \\
\downarrow & & \downarrow {\scriptstyle t_* X} \\
1 & \underset{t}{\rightarrowtail} & \Phi.
\end{array}
\tag{3.1.2}
$$

Let $\eta : X \rightarrowtail X^+$ be the indicated top horizontal map; we call this the *cofibrant partial map classifier* of X. By a *cofibrant partial map* (from an object Z) into X we mean a span $(c, x) : Z \leftarrowtail C \to X$ with a cofibration on the left. The object X^+ is a *classifying type* for such cofibrant partial maps, in the sense that it has the following universal property.

Proposition 3.2 *Let* $\eta : X \rightarrowtail X^+$ *be as defined in* (3.1.2).

1. *The map* $\eta : X \rightarrowtail X^+$ *is a cofibration.*
2. *For any object* Z *and any partial map* $(c, x) : Z \leftarrowtail C \to X$, *with* $c : C \rightarrowtail Z$ *a cofibration, there is a unique* $\chi : Z \to X^+$ *fitting into a pullback square as follows.*

$$
\begin{array}{ccc}
C & \overset{x}{\longrightarrow} & X \\
{\scriptstyle c}\downarrow & & \downarrow {\scriptstyle \eta} \\
Z & \underset{\chi}{\longrightarrow} & X^+
\end{array}
$$

The map $\chi : Z \to X^+$ *is said to* classify *the partial map*

$$(c, x) : Z \leftarrowtail C \to X .$$

Proof The map $\eta : X \rightarrowtail X^+$ is a cofibration, since it is a pullback of the universal cofibration $t : 1 \rightarrowtail \Phi$. Observe that

$$(\eta, 1_X) : X^+ \leftarrowtail X \rightarrow X$$

is therefore a cofibrant partial map into X. The second statement is just the universal property of X^+ as a polynomial, which can be read off from the description (3.1.1) (see [7], prop. 7). □

Proposition 3.3 *The pointed endofunctor* $\eta_X : X \rightarrowtail X^+$ *has a natural multiplication* $\mu_X : X^{++} \rightarrow X^+$ *making it a monad.*

Proof Since the cofibrations are closed under composition, the monad structure on X^+ follows as in [7], Lemma 5. Explicitly, μ_X is determined by Proposition 3.2 as the unique map making the following a pullback diagram.

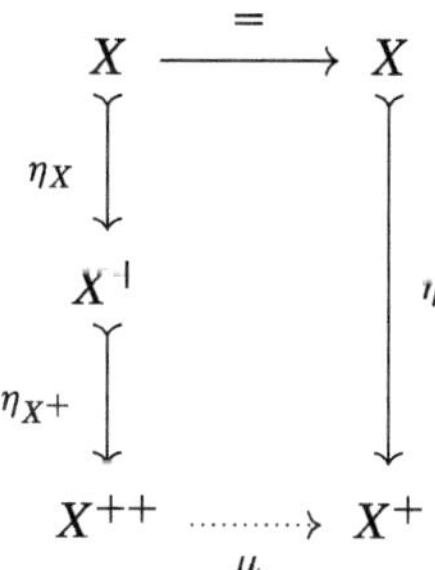

□

3.2   Relative Partial Map Classifier

For any object $X \in \mathcal{E}$ the pullback functor

$$X^* : \mathcal{E} \rightarrow \mathcal{E}/X \, ,$$

taking any A to the (say) first projection $X \times A \rightarrow X$, not only preserves the subobject classifier Ω, but also the cofibration classifier $\Phi \hookrightarrow \Omega$, where a map in $\mathcal{E}/X$ is defined to be a cofibration just if it is one in $\mathcal{E}$ (under the forgetful functor $\mathcal{E}/X \rightarrow \mathcal{E}$). Thus in $\mathcal{E}/X$ we can define the *(relative) cofibration classifier* to be the map

$$X^* t : X^* 1 \longrightarrow X^* \Phi \quad \text{over } X \, ,$$

which we may also write $t_X : 1_X \to \Phi_X$. Like $t : 1 \to \Phi$, this map determines a polynomial endofunctor

$$(-)^{+x} : \mathcal{E}/X \longrightarrow \mathcal{E}/X ,$$

which commutes (up to natural isomorphism) with $(-)^+ : \mathcal{E} \to \mathcal{E}$ and $X^* : \mathcal{E} \to \mathcal{E}/X$ in the expected way, namely:

$$
\begin{array}{ccc}
\mathcal{E}/X & \xrightarrow{\ +x\ } & \mathcal{E}/X \\[4pt]
X^* \uparrow & & \uparrow X^* \\[4pt]
\mathcal{E} & \xrightarrow{\ +\ } & \mathcal{E}
\end{array}
\tag{3.2.1}
$$

The endofunctor $+_X$ is also pointed $\eta_Y : Y \to Y^+$ and has a natural monad multiplication $\mu_Y : Y^{++} \to Y^+$, for any $Y \to X$, for the same reason that $+$ has this structure. Summarizing, we may say:

Proposition 3.4 *The polynomial monad* $(-)^+ : \mathcal{E} \to \mathcal{E}$ *of cofibrant partial elements is indexed (or fibered) over* $\mathcal{E}$.

Definition 3.5 A $+$-*algebra* in $\mathcal{E}$ is an algebra for the pointed endofunctor $(-)^+ : \mathcal{E} \to \mathcal{E}$. Explicitly, a $+$-algebra is an object A together with a retraction $\alpha : A^+ \to A$ of the unit $\eta_A : A \to A^+$. Algebras for the *monad* $(+, \eta, \mu)$ will be referred to explicitly as $(+, \eta, \mu)$-*algebras*, or $+$-*monad algebras*.

A *relative* $+$-*algebra* in $\mathcal{E}$ is a map $A \to X$, together with an algebra structure over the codomain X for the pointed endofunctor

$$(-)^{+x} : \mathcal{E}/X \longrightarrow \mathcal{E}/X .$$

3.3 The Cofibration Weak Factorization System

The following proposition generalizes one in [20].

Proposition 3.6 *There is an (algebraic) weak factorization system on* $\mathcal{E}$ *with the cofibrations as the left class, and as the right class the underlying maps of relative*

+-algebras. Thus a right map is one $p : A \to X$ for which there is a retract $\alpha : A' \to A$ over X of the canonical map $\eta : A \to A'$,

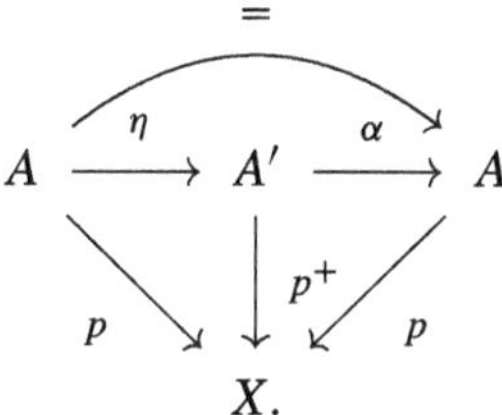

(Note that the domain of $p^+ : A' \to X$ is not A^+, unless of course $X = 1$.)

Proof The factorization of a map $f : Y \to X$ is given by applying the relative $+$-functor over the codomain,

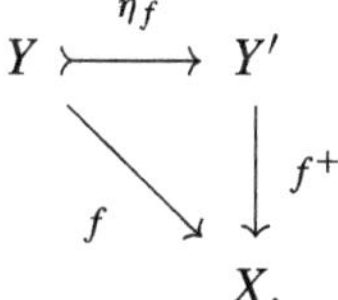

We know by Proposition 3.2 that the unit η_f is always a cofibration, and since f^+ is the free monad algebra for the relative $+$-monad, it is in particular a $+$-algebra.

For the lifting condition, consider a cofibration $c : B \rightarrowtail C$, and a right map $p : A \to X$ with $+$-algebra structure map $\alpha : A' \to A$ over X, and a commutative square as indicated below.

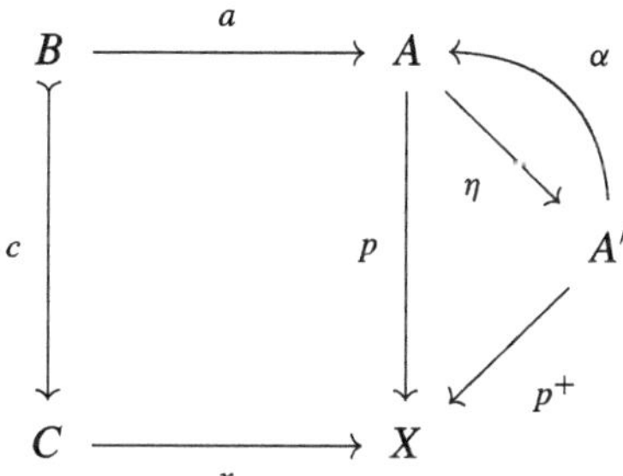

Viewed in the slice category over X, we have

and we seek an extension d of a along c, as indicated. (Note that we are writing A^+ for the map $p^+ : A' \to A$ regarded as an object over X, and similarly C for $x : C \to X$ and B for $xc : B \to X$ and A for $p : A \to X$.) Since $(c, a) : B \hookleftarrow C \to A$ is a cofibrant partial map into A, by the universal property of $\eta : A \rightarrowtail A^+$ (Proposition 3.2) there is a unique classifying map $\chi : C \to A^+$ (over X) making a pullback square,

$$
\begin{array}{ccc}
B & \xrightarrow{\ a\ } & A \\
{\scriptstyle c}\downarrow & & \downarrow{\scriptstyle \eta} \\
C & \dashrightarrow[\chi] & A^+.
\end{array}
$$

We can set $d := \alpha \circ \chi : C \to A$ to obtain the required diagonal filler, since $dc = \alpha \chi c = \alpha \eta a = a$, because the square commutes, and α is a retract of η.

The closure of the cofibrations under retracts follows from their classification by a universal object $t : 1 \rightarrowtail \Phi$, and the closure of the right maps under retracts follows from their being the algebras for a pointed endofunctor underlying a monad (cf. [65]). Algebraicity of this weak factorization system follows from the fact that $(-)^+$ is a fibered monad. $\qquad\square$

Summarizing, we have an algebraic weak factorization system $(\mathcal{C}, \mathcal{C}^{\pitchfork})$ on the category $\mathcal{E}$, where:

$$\mathcal{C} = \text{the cofibrations}$$

$$\mathcal{C}^{\pitchfork} = \text{the maps underlying relative} + -\text{algebras}$$

We shall call this the *cofibration weak factorization system*. The right maps will be called *trivial fibrations*, and the class of all such denoted

$$\mathsf{TFib} := \mathcal{C}^{\pitchfork}.$$

The cofibration algebraic weak factorization system is a generalization of one defined in [20] and mentioned in [35].

3.4 Uniform Filling Structure

It will be useful to relate relative $+$-algebra structure to the more familiar diagonal filling condition of cofibrantly generated weak factorization systems, and specifically the special ones occurring in [28] under the name *uniform filling structure* (this notion is also closely related to that of an *algebraic weak factorization system*, cf. [36, 64]). For the purpose of such comparison, we assume in the remainder of this chapter that $\mathcal{E} = \mathsf{Set}^{\mathbb{C}^{op}}$ is a topos of presheaves on a small index category

$\mathbb{C}$, and we write the representable functors in the form $\mathsf{I} = \mathsf{y}(i)$, $\mathsf{J} = \mathsf{y}(j)$, ... for $i, j, \ldots$ in $\mathbb{C}$.

Consider a generating sub*set* of cofibrations consisting of those with representable codomain $c : C \rightarrowtail \mathsf{I}$, and call these the *basic cofibrations*.

$$\mathsf{BCof} = \{c : C \rightarrowtail \mathsf{I} \mid c \in C,\ \mathsf{I} \text{ representable}\}. \tag{3.4.1}$$

Proposition 3.7 *For any object X in $\mathcal{E}$ the following are equivalent:*

1. *X admits a $+$-algebra structure: a retraction $\alpha : X^+ \to X$ of the unit $\eta : X \to X^+$.*
2. *$X \to 1$ is a trivial fibration: it has the right lifting property with respect to all cofibrations,*

$$C \pitchfork X.$$

3. *X admits a uniform filling structure: for each basic cofibration $c : C \rightarrowtail \mathsf{I}$ and map $x : C \to X$ there is given an extension $j(c, x)$,*

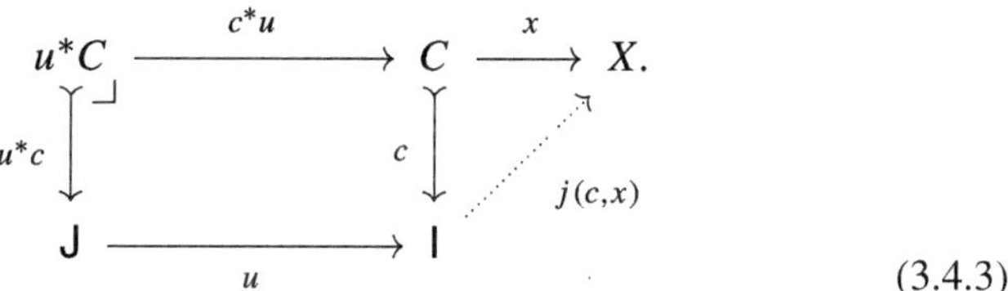

$$\tag{3.4.2}$$

and the choice is uniform in I in the following sense.

   Given any map of representables $u : \mathsf{J} \to \mathsf{I}$, the pullback $u^*c : u^*C \rightarrowtail \mathsf{J}$, which is again a basic cofibration, fits into a commutative diagram of the form

$$\tag{3.4.3}$$

For the pair $(u^*c, x \circ c^*u)$ in (3.4.3), the chosen extension $j(u^*c, x \circ c^*u) : \mathsf{J} \to X$, is required to be equal to $j(c, x) \circ u$,

$$j(u^*c, x \circ c^*u) = j(c, x) \circ u. \tag{3.4.4}$$

Proof Let (X, α) be a $+$-*algebra* and suppose given the span (c, x) as below, with c a cofibration.

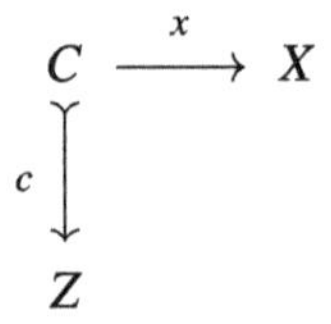

$$
\begin{array}{ccc}
C & \xrightarrow{\;\;x\;\;} & X \\
{\scriptstyle c}\downarrow & & \\
Z & &
\end{array}
$$

Let $\chi(c, x) : Z \to X^+$ be the classifying map of the cofibrant partial map $(c, x) :$ $Z \leftarrowtail C \to X$, so that we have a pullback square as follows.

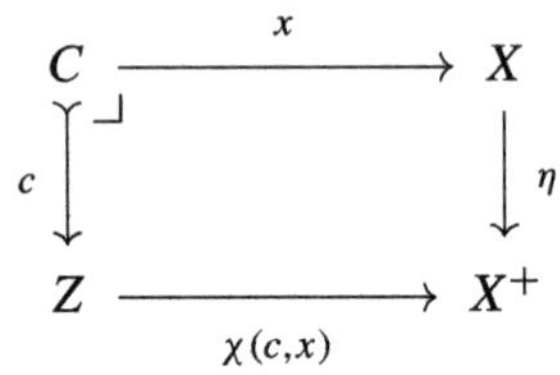

$$
\begin{array}{ccc}
C & \xrightarrow{\;\;x\;\;} & X \\
{\scriptstyle c}\downarrow & & \downarrow{\scriptstyle \eta} \\
Z & \xrightarrow[\chi(c,x)]{} & X^+
\end{array}
\tag{3.4.5}
$$

Then set

$$
j = \alpha \circ \chi(c, x) : Z \to X
\tag{3.4.6}
$$

to get a filler,

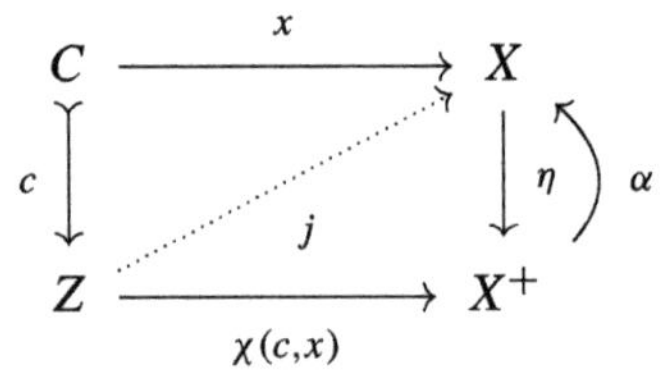

$$
\begin{array}{ccc}
C & \xrightarrow{\;\;x\;\;} & X \\
{\scriptstyle c}\downarrow & \nearrow{\scriptstyle j} & \downarrow{\scriptstyle \eta}\;\;\;\Big)\,{\scriptstyle \alpha} \\
Z & \xrightarrow[\chi(c,x)]{} & X^+
\end{array}
\tag{3.4.7}
$$

since

$$
j \circ c = \alpha \circ \chi(c, x) \circ c = \alpha \circ \eta \circ x = x.
$$

Thus (1) implies (2). To see that it also implies (3), observe that in the case where $Z = \mathsf{I}$ representable, and we specify, in (3.4.6), that

$$
j(c, x) = \alpha \circ \chi(c, x) : \mathsf{I} \to X,
\tag{3.4.8}
$$

the assignment is then natural in I. Indeed, given any $u : \mathsf{J} \to \mathsf{I}$, we have

$$
j(c', xu') = \alpha \circ \chi(c', xu') = \alpha \circ \chi(c, x) \circ u = j(c, x)u,
\tag{3.4.9}
$$

by the uniqueness of the classifying maps.

It is clear that (2) implies (1), since if $C \pitchfork X$ then we can take as an algebra structure $\alpha : X^+ \to X$ any filler for the universal span

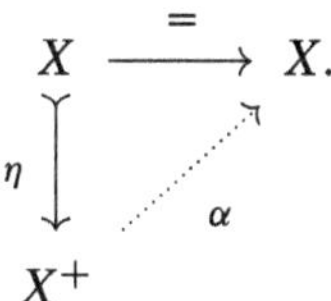

To see that (3) implies (1), suppose that X has a uniform filling structure j and we want to define an algebra structure $\alpha : X^+ \to X$. By Yoneda, for every $y : I \to X^+$ we need a map $\alpha(y) : I \to X$, naturally in I, in the sense that for any $u : J \to I$, we have

$$\alpha(yu) = \alpha(y)u. \tag{3.4.10}$$

Moreover, to ensure that $\alpha\eta = 1_X$, for any $x : I \to X$ we must have $\alpha(\eta \circ x) = x$. So take $y : I \to X^+$ and let

$$\alpha(y) = j(y^*\eta, y'),$$

as indicated on the right below.

$$
\begin{array}{ccccc}
u^*C & \xrightarrow{\;u'\;} & C & \xrightarrow{\quad y'\quad} & X. \\
\Big\downarrow{\scriptstyle u^*y^*\eta} & & \Big\downarrow{\scriptstyle y^*\eta} & \overset{j(y^*\eta,y')}{\nearrow} & \Big\downarrow{\scriptstyle \eta} \\
J & \xrightarrow[\;u\;]{} & I & \xrightarrow[\;y\;]{} & X^+
\end{array}
\tag{3.4.11}
$$

Then for any $u : J \to I$, we indeed have

$$\alpha(yu) = j\big((yu)^*\eta, \, y'u'\big) = j(y^*\eta, y') \circ u = \alpha(y)u,$$

by the uniformity of j. Finally, if $y = \eta \circ x$ for some $x : I \to X$ then

$$\alpha(\eta x) = j\big((\eta x)^*\eta, (\eta x)'\big) = j(1_X, x) = x,$$

because the defining diagram for $\alpha(\eta x)$, i.e. the one on the right in (3.4.11), then factors as

$$(3.4.12)$$

and the only possible extension $j(1_X, x)$ for the span $(1_I, x)$ is x itself. $\square$

Remark 3.8 Observe that the uniformity condition (3) can be extended to the *class of all* cofibrations, in the form:

4. X admits a *(large) uniform filling structure:* for each cofibration $c : C \rightarrowtail Z$ and map $x : C \to X$ there is given an extension $j(c, x)$,

$$(3.4.13)$$

and the choice is *uniform in Z* in the following sense: Given any map $u : Y \to Z$, the pullback $u^*c : u^*C \rightarrowtail Y$, which is again a cofibration, fits into a commutative diagram of the form

$$(3.4.14)$$

For the pair $(u^*c, x \circ c^*u)$ in (3.4.14), the chosen extension $j(u^*c, x \circ c^*u) : Y \to X$, is required to be equal to $j(c, x) \circ u$,

$$j(u^*c, x \circ c^*u) = j(c, x) \circ u. \qquad (3.4.15)$$

Indeed, the proof that (1) implies (2) and (3) works just as well to infer (4), which in turn implies (2) and (3) as special cases. The equivalence of (1) and (2) of Proposition 3.7 and condition (4) therefore holds for arbitrary (not necessarily presheaf) toposes.

The relative version of the foregoing is entirely analogous, since the +-monad is fibered over $\mathcal{E}$ in the sense of diagram (3.2.1). We therefore omit the entirely analogous proof of the following (which can also be derived from Proposition 3.7 by relativizing to the slice $\mathcal{E}/Y$).

Proposition 3.9 *For any map* $f : X \to Y$ *in* $\mathcal{E}$ *the following are equivalent:*

1. $f : X \to Y$ *admits a* relative $+$*-algebra structure over* Y, *i.e. there is a retraction* $\alpha : X' \to X$ *over* Y *of the unit* $\eta : X \to X'$, *where* $f^{+} : X' \to Y$ *is the result of the relative* $+$*-functor applied to* f, *as in Definition 3.5.*
2. $f : X \to Y$ *is a* trivial fibration,

$$C \pitchfork f.$$

3. $f : X \to Y$ *admits a (small)* uniform filling structure: *for each basic cofibration* $c : C \rightarrowtail I$ *and maps* $x : C \to X$ *and* $y : I \to Y$ *making the square below commute, there is given a diagonal filler* $j(c, x, y)$,

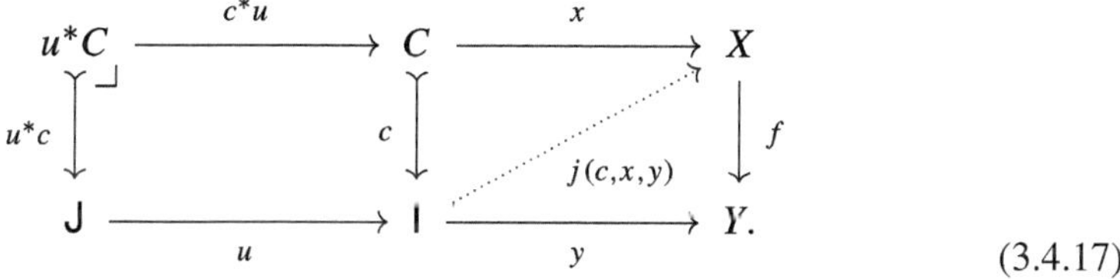

$$\tag{3.4.16}$$

and the choice is uniform in I *in the following sense: given any map* $u : J \to I$, *the pullback* $u^{*}c : u^{*}C \rightarrowtail J$ *is again a basic cofibration and fits into a commutative diagram of the form*

$$\tag{3.4.17}$$

For the evident triple $(u^{*}c, x \circ c^{*}u, y \circ u)$ *in (3.4.17) the chosen diagonal filler*

$$j(u^{*}c, x \circ c^{*}u, y \circ u) : J \to X$$

is equal to $j(c, x, y) \circ u$,

$$j(u^{*}c, x \circ c^{*}u, y \circ u) = j(c, x, y) \circ u. \tag{3.4.18}$$

And again, a large version of (3) with arbitrary cofibrations $c : C \rightarrowtail Z$ *is equivalent to (1)–(3), so the relative version holds also for arbitrary (not necessarily presheaf) toposes.*

Finally, let us collect some basic facts about trivial fibrations that will be needed later: they have sections, they are closed under composition and retracts, and they are closed under pullback and pushforward along all maps.

Corollary 3.10

1. *Every trivial fibration $A \to X$ has a section $s : X \to A$.*
2. *If $a : A \to X$ is a trivial fibration and $b : B \to A$ is a trivial fibration, then $a \circ b : B \to X$ is a trivial fibration.*
3. *If $a : A \to X$ is a trivial fibration and $a' : A' \to X'$ is a retract of a in the arrow category, then a' is a trivial fibration.*
4. *For any map $f : X \to Y$ and any trivial fibration $B \to Y$, the pullback $f^*B \to X$ is a trivial fibration.*
5. *For any map $f : X \to Y$ and any trivial fibration $A \to X$, the pushforward $f_*A \to Y$ is a trivial fibration.*

Proof (1) holds because all objects are cofibrant by (C0). (5) is a consequence of (C3), stability of cofibrations under pullback, by a standard argument using the adjunction $f^* \dashv f_*$. The rest hold for the right maps in any weak factorization system. $\qquad\qquad\square$

Remark 3.11 The structured notion of trivial fibration, vis. relative $+$-algebra, can also be shown algebraically (i.e. not using Proposition 3.9) to be closed under composition and retracts and preserved by pullback and pushforward. We consider the case of pushforward as an example. Thus consider the following situation with $A \to X$ a $+$-algebra with structure α, as indicated.

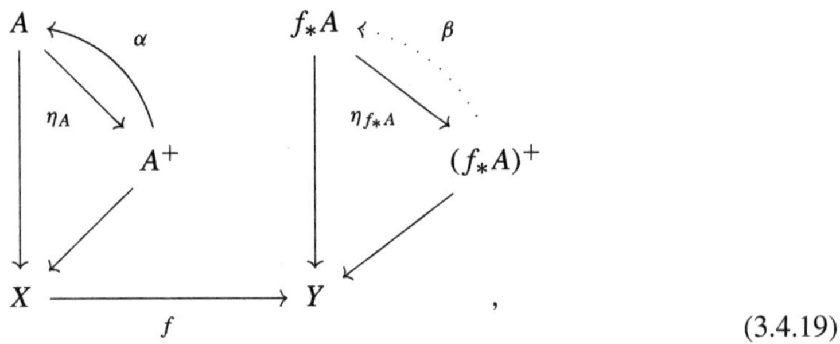

$$(3.4.19)$$

A $+$-algebra structure for $f_*A \to Y$ would be a retract $\beta : (f_*A)^+ \to f_*A$ of $\eta_{f_*A} : f_*A \to (f_*A)^+$ over Y, which corresponds under $f^* \dashv f_*$ to a map $\tilde{\beta} : f^*((f_*A)^+) \to A$ over X with

$$\tilde{\beta} \circ f^* \eta_{f_*A} = \epsilon_A$$

as indicated below.

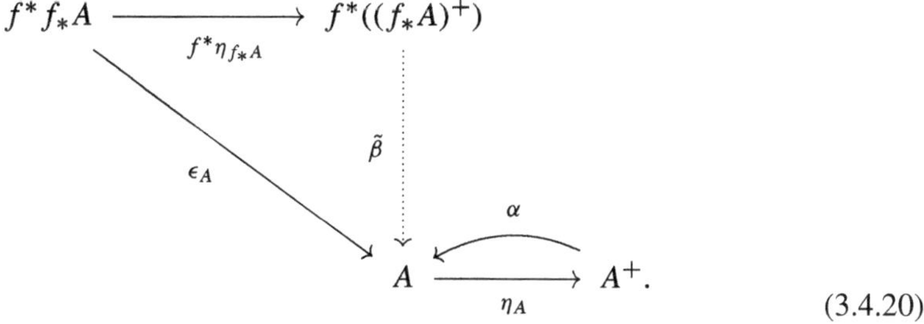

$$\tag{3.4.20}$$

But since pullback f^* commutes with $+$, there is a canonical iso $c : f^*((f_*A)^+) \cong (f^*f_*A)^+$ with $c \circ f^*\eta_{f_*A} = \eta_{f^*f_*A}$. So we can set $\tilde{\beta} := \alpha \circ (\epsilon_A)^+ \circ c$.

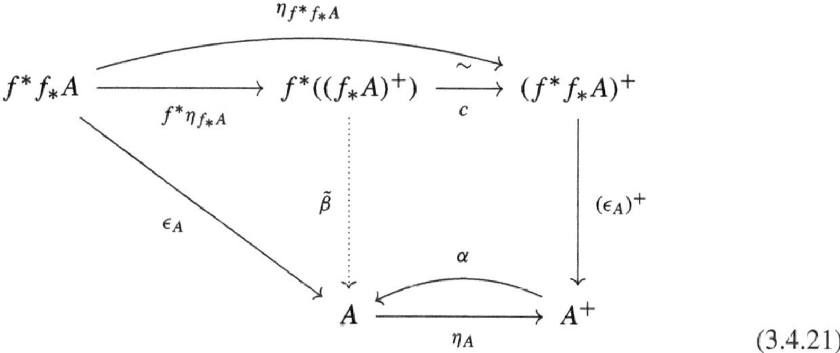

$$\tag{3.4.21}$$

Chapter 4
The Fibration Weak Factorization System

Returning from the elementary case to the particular presheaf topos cSet of cubical sets, we now specify a second weak factorization system, with a restricted class of "trivial" cofibrations on the left, and an expanded class of right maps, the *fibrations*. As explained in the introduction, we first recall from [35] what we shall call the "biased" notion of fibration, before generalizing the endpoints $\delta_0, \delta_1 \,:\, 1 \to I$ to arbitrary points $\delta : 1 \to I$ in the "unbiased" version appropriate to the more general Cartesian setting. The two versions are equivalent in the presence of *connections*

$$\vee, \wedge : I \times I \longrightarrow I$$

on the cubes, which are used in [66] to determine a model structure with biased fibrations. In [14] it is shown that the biased fibrations of op.cit. agree with those specified in the "logical style" of [28, 60]. This is not the case, however, for the purely Cartesian case, without connections, which are *not* assumed in the category $\square$ of Cartesian cubes. The unbiased approach will amount to adding further weak equivalences (fewer fibrations, therefore more *trivial* cofibrations). In this setting, the methods of [66] no longer apply; we must therefore find new proofs of several basic results, including notably the Frobenius condition.

4.1 Partial Box Filling (Biased Version)

The *generating biased trivial cofibrations* are all maps of the form

$$c \otimes \delta_\epsilon : D \rightarrowtail Z \times I, \tag{4.1.1}$$

where:

1. $c : C \rightarrowtail Z$ is an arbitrary cofibration,
2. $\delta_\epsilon : 1 \to I$ is one of the two *endpoint inclusions*, for $\epsilon = 0, 1$.
3. $c \otimes \delta_\epsilon$ is the *pushout-product* indicated in the following diagram.

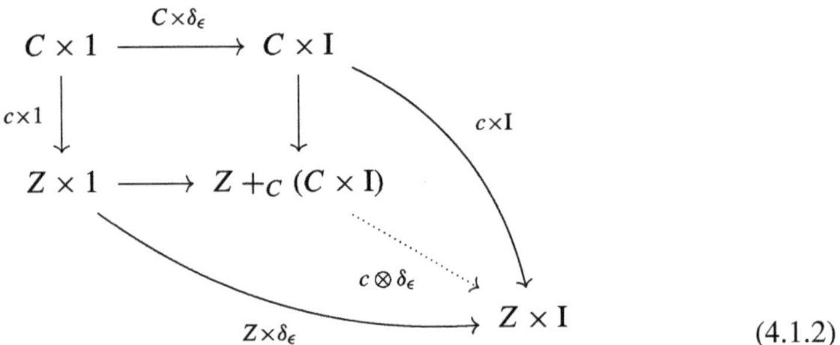

$$(4.1.2)$$

4. $D = Z +_C (C \times I)$ is the indicated domain of the map $c \otimes \delta_\epsilon$.

In order to ensure that such maps are indeed cofibrations, we henceforth assume two further axioms in addition to (C1)–(C4) from Definition 3.1:

(C5) The endpoint inclusions $\delta_\epsilon : 1 \to I$ are cofibrations, for $\epsilon = 0, 1$.
(C6) The cofibrations are closed under joins $A \vee B \rightarrowtail C$ of subobjects $A, B \rightarrowtail C$ of any object C.

Remark 4.1 Note that since $\delta_0 : 1 \to I$ and $\delta_1 : 1 \to I$ are disjoint, by (C5) and stability under pullbacks we have that $0 \to 1$ is a cofibration, so by stability again $0 \to A$ is always a cofibration. Thus (C0) is no longer required. From (C6) it follows that cofibrations are closed under pushout-products $a \otimes b$ in the arrow category. It also then follows from (C5) that the boundary $\partial : 1 + 1 \to I$ is a cofibration.

4.2   Fibrations (Biased Version)

Now let

$$C \otimes \delta_\epsilon = \{ c \otimes \delta_\epsilon : D \rightarrowtail Z \times I \mid c \in C,\ \epsilon = 0, 1 \}$$

be the class of all generating biased trivial cofibrations. The *biased fibrations* are defined to be the right class of these maps,

$$(C \otimes \delta_\epsilon)^{\pitchfork} = \mathcal{F}.$$

Thus a map $f : Y \to X$ is a biased fibration just if for every commutative square of the form

$$
\begin{array}{ccc}
Z +_C (C \times I) & \longrightarrow & Y \\
{\scriptstyle c \otimes \delta_\epsilon} \downarrow & {\scriptstyle j} \nearrow & \downarrow {\scriptstyle f} \\
Z \times I & \longrightarrow & X
\end{array}
\qquad (4.2.1)
$$

with a generating biased trivial cofibration on the left, there is a diagonal filler j as indicated.

To relate this notion of fibration to the cofibration weak factorization system, fix any map $u : A \to B$, and recall (e.g. from [47, 65]) that the pushout-product with u is a functor on the arrow category

$$
(-) \otimes u : \mathsf{cSet}^2 \to \mathsf{cSet}^2 .
$$

This functor has a right adjoint, the *pullback-hom*, which for a map $f : Y \to X$ we shall write as

$$
(u \Rightarrow f) : Y^B \longrightarrow (X^B \times_{X^A} Y^A) .
$$

The pullback-hom is determined as indicated in the following diagram.

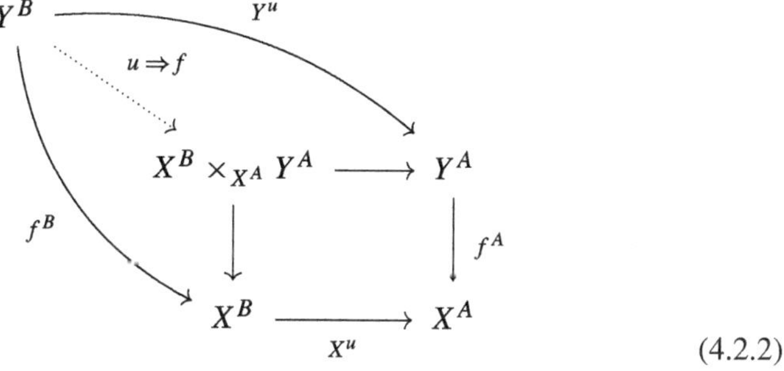

$$
\qquad (4.2.2)
$$

The $\otimes \dashv \Rightarrow$ adjunction on the arrow category has the following useful relation to weak factorization systems (cf. [35, 47, 65]), where for any maps $a : A \to B$ and $f : Y \to X$ we write

$$
a \pitchfork f
$$

to mean that for every solid square of the form

$$
\begin{array}{ccc}
A & \longrightarrow & Y \\
a \downarrow & \nearrow_{\;j} & \downarrow f \\
B & \longrightarrow & X
\end{array}
\tag{4.2.3}
$$

there exists a diagonal filler j as indicated.

Lemma 4.2 *For any maps $a : A_0 \to A_1$, $b : B_0 \to B_1$, $c : C_0 \to C_1$ in* cSet,

$$
(a \otimes b) \pitchfork c \quad \textit{iff} \quad a \pitchfork (b \Rightarrow c) .
$$

The following is now a direct corollary.

Proposition 4.3 *An object X is fibrant if and only if both of the endpoint projections $X^I \to X$ from the pathspace are trivial fibrations. More generally, a map $f : Y \to X$ is a fibration just if both of the maps*

$$
(\delta_\epsilon \Rightarrow f) : Y^I \to X^I \times_X Y
$$

are trivial fibrations (for $\epsilon = 0, 1$).

4.3   Fibration Structure (Biased Version)

The $\otimes \dashv \Rightarrow$ adjunction determines the fibrations in terms of the trivial fibrations, which in turn can be determined by *uniform* lifting against a *small category* consisting of basic cofibrations and pullback squares between them, by Proposition 3.9. The fibrations are similarly determined by *uniform* lifting against the *small category* of basic, biased trivial cofibrations, consisting of all those $c \otimes \delta_\epsilon$ in $C \otimes \delta_\epsilon$ where $c : C \rightarrowtail I^n$ is a *basic* cofibration, i.e. one with representable codomain. Thus the set of *basic biased trivial cofibrations* is

$$
\mathsf{BCof} \otimes \delta_\epsilon = \{ c \otimes \delta_\epsilon : B \rightarrowtail I^{n+1} \mid c : C \rightarrowtail I^n, \ \epsilon = 0, 1, \ n \geq 0 \},
\tag{4.3.1}
$$

where the pushout-product $c \otimes \delta_\epsilon$ now takes the simpler form

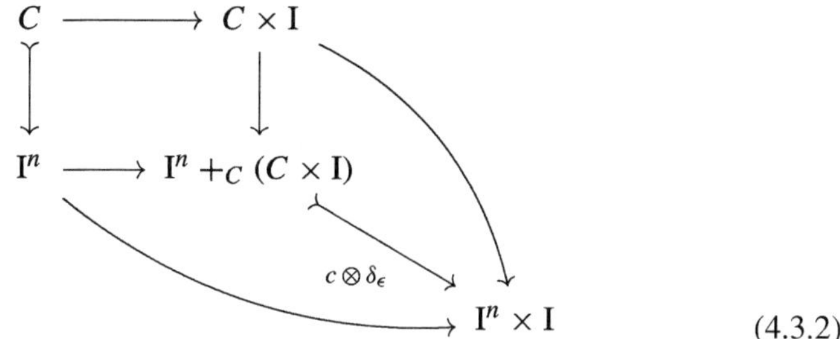

$$(4.3.2)$$

for a basic cofibration $c : C \rightarrowtail I^n$, an endpoint $\delta_\epsilon : 1 \to I$, and with domain $B = (I^n +_C (C \times I))$. These subobjects $B \rightarrowtail I^{n+1}$ can be seen geometrically as generalized open box inclusions.

For any map $f : Y \to X$ a *uniform, biased fibration structure* on f is a choice of diagonal fillers $j_\epsilon(c, x, y)$,

$$
\begin{array}{ccc}
I^n +_C (C \times I) & \xrightarrow{\ \ x\ \ } & X \\
{\scriptstyle c \otimes \delta_\epsilon} \downarrow & {\scriptstyle j_\epsilon(c,x,y)} & \downarrow {\scriptstyle f} \\
I^n \times I & \xrightarrow[\ \ y\ \ ]{} & Y,
\end{array}
$$

$$(4.3.3)$$

for each basic biased trivial cofibration $c \otimes \delta_\epsilon : B = (I^n +_C (C \times I)) \rightarrowtail I^{n+1}$ and maps $x : B \to X$ and $y : I^{n+1} \to Y$, which is *uniform in* I^n in the following sense: Given any cubical map $u : I^m \to I^n$, the pullback $u^*c : u^*C \rightarrowtail I^m$ of $c : C \rightarrowtail I^n$ along u determines another basic biased trivial cofibration

$$u^*c \otimes \delta_\epsilon : B' = (I^m +_{u^*C} (u^*C \times I)) \rightarrowtail I^{m+1},$$

which fits into a commutative diagram of the form

$$
\begin{array}{ccccc}
I^m +_{u^*C} (u^*C \times I) & \xrightarrow{\ \ (u \times I)'\ \ } & I^n +_C (C \times I) & \xrightarrow{\ \ x\ \ } & X \\
{\scriptstyle u^*c \otimes \delta_\epsilon} \downarrow & & {\scriptstyle c \otimes \delta_\epsilon} \downarrow \quad {\scriptstyle j_\epsilon(c,x,y)} & & \downarrow {\scriptstyle f} \\
I^m \times I & \xrightarrow[\ \ u \times I\ \ ]{} & I^n \times I & \xrightarrow[\ \ y\ \ ]{} & Y,
\end{array}
$$

$$(4.3.4)$$

by applying the functor $(-) \otimes \delta_\epsilon$ to the pullback square relating u^*c to c. For the outer rectangle in (4.3.4) there is then a chosen diagonal filler

$$j_\epsilon(u^*c, x \cup (u \times I)', y \circ (u \times I)) : I^m \times I \to X ,$$

and for this map we require that

$$j_\epsilon(u^*c, x \circ (u \times \mathrm{I})', y \circ (u \times \mathrm{I})) = j_\epsilon(c, x, y) \circ (u \times \mathrm{I}). \tag{4.3.5}$$

This can be seen to be a reformulation of the logical specification given in [28] (see [14]).

Definition 4.4 A *uniform, biased fibration structure* on a map $f : Y \to X$ is a choice of fillers $j_\epsilon(c, x, y)$ as in (4.3.3) satisfying (4.3.5) for all maps $u : \mathrm{I}^m \to \mathrm{I}^n$.

Finally, we have the analogue of Proposition 3.7 for fibrant objects. The analogous statement of Proposition 3.9 for fibrations is omitted, as is the entirely analogous proof.

Corollary 4.5 *For any object X in* cSet *the following are equivalent:*

1. *X is* biased fibrant, *in the sense that every map $D \to X$ from the domain of a generating biased trivial cofibration $D \rightarrowtail Z \times \mathrm{I}$ extends to a total map $Z \times \mathrm{I} \to X$,*

$$C \otimes \delta_\epsilon \pitchfork X .$$

2. *The canonical maps $(\delta_\epsilon \Rightarrow X) : X^I \to X$ are trivial fibrations.*
3. *$X \to 1$ admits a* uniform biased fibration structure. *Explicitly, for each basic biased trivial cofibration $c \otimes \delta_\epsilon : B \rightarrowtail \mathrm{I}^{n+1}$ and map $x : B \to X$, there is given an extension $j_\epsilon(c, x)$,*

$$\begin{array}{ccc} B & \xrightarrow{\;\;x\;\;} & X, \\ {\scriptstyle c \otimes \delta_\epsilon} \big\downarrow & \nearrow & \\ \mathrm{I}^{n+1} & {\scriptstyle j_\epsilon(c,x)} & \end{array} \tag{4.3.6}$$

and, moreover, the choice is uniform in I^n *in the following sense: Given any cubical map $u : \mathrm{I}^m \to \mathrm{I}^n$, the pullback $u^*c \otimes \delta_\epsilon : B' \rightarrowtail \mathrm{I}^m \times \mathrm{I}$ fits into a commutative diagram of the form*

$$\begin{array}{ccccc} B' & \xrightarrow{\;\;(u \times \mathrm{I})'\;\;} & B & \xrightarrow{\;\;x\;\;} & X. \\ {\scriptstyle u^*c \otimes \delta_\epsilon} \big\downarrow & & {\scriptstyle c \otimes \delta_\epsilon} \big\downarrow & \nearrow & \\ \mathrm{I}^m \times \mathrm{I} & \xrightarrow[\;\;u \times \mathrm{I}\;\;]{} & \mathrm{I}^n \times \mathrm{I} & {\scriptstyle j(c,x)} & \end{array} \tag{4.3.7}$$

For the pair $(u^*c \otimes \delta_\epsilon, x \circ (u \times \mathrm{I})')$ in (4.3.7) the chosen extension

$$j(u^*c \otimes \delta_\epsilon, x \circ (u \times \mathrm{I})') : \mathrm{I}^m \times \mathrm{I} \to X$$

is equal to $j(c, x) \circ (u \times \mathrm{I})$,

$$j(u^*c \otimes \delta_\epsilon, x \circ (u \times \mathrm{I})') = j(c, x)(u \times \mathrm{I}).$$ (4.3.8)

4.4   Partial Box Filling (Unbiased Version)

Rather than building a weak factorization system based on the foregoing notion of biased fibration (as is done in [35]), we shall first eliminate the "bias" with respect to the endpoints $\delta_\epsilon : 1 \rightarrow \mathrm{I}$, for $\epsilon = 0, 1$. This will have the effect of adding more trivial cofibrations, and thus more weak equivalences, to our model structure. Consider first the simple path-lifting condition for a map $f : Y \rightarrow X$, which is a special case of (4.2.1) with $c = \,! : 0 \rightarrowtail 1$, so that $! \otimes \delta_\epsilon = \delta_\epsilon$.

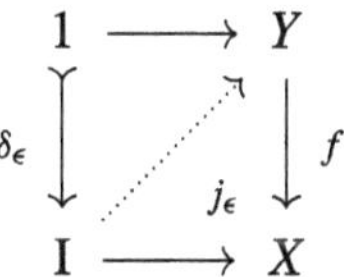

In topological spaces, for instance, rather than requiring lifts j_ϵ for each of the endpoints $\epsilon = 0, 1$ of the real interval $\mathrm{I} = [0, 1]$, one could equivalently require there to be a lift j_i *for each point* $i : 1 \rightarrow \mathrm{I}$. Such "unbiased path-lifting" can be formulated in cSet by introducing a "generic point" $\delta : 1 \rightarrow \mathrm{I}$ by passing to $\mathsf{cSet}_{/\mathrm{I}}$ via the pullback functor $\mathrm{I}^* : \mathsf{cSet} \rightarrow \mathsf{cSet}_{/\mathrm{I}}$ and then requiring path-lifting for I^*f with respect to $\delta : \mathrm{I} \rightarrow \mathrm{I} \times \mathrm{I}$, regarded as a map $\delta : 1 \rightarrow \mathrm{I}^*\mathrm{I}$ in $\mathsf{cSet}_{/\mathrm{I}}$. We shall therefore define f to be an unbiased fibration just if I^*f is a δ-biased fibration for the generic point δ. The following specification implements that idea, while also adding cofibrant partiality, as in the biased case.

We begin by replacing axiom (C5) with the following, stronger assumption, which will be assumed henceforth.

(C7)   The diagonal map $\delta : \mathrm{I} \rightarrow \mathrm{I} \times \mathrm{I}$ of the interval I is a cofibration.

The unbiased notion of a fibration for cSet is now as follows.

Definition 4.6 (Unbiased Fibration)   Let $\delta : \mathrm{I} \rightarrow \mathrm{I} \times \mathrm{I}$ be the diagonal map.

1. An object X is *unbiased fibrant* if the map

$$(\delta \Rightarrow X) = \langle \mathsf{eval}, p_2 \rangle : X^{\mathrm{I}} \times \mathrm{I} \rightarrow X \times \mathrm{I}$$

is a trivial fibration.

2. A map $f : Y \to X$ is an *unbiased fibration* if the map

$$(\delta \Rightarrow f) = \langle f^I \times I, \langle \mathsf{eval}, p_2 \rangle \rangle : Y^I \times I \to (X^I \times I) \times_{(X \times I)} (Y \times I)$$

is a trivial fibration.

Let us (temporarily) write $\mathbb{I} = I^*I$ for the pulled-back interval in the slice category $\mathsf{cSet}/_I$, so that the generic point is written $\delta : 1 \to \mathbb{I}$. Condition (1) above (which of course is a special case of (2)) then says that evaluation at the generic point $\delta : 1 \to \mathbb{I}$, the map $(I^*X)^\delta : (I^*X)^{\mathbb{I}} \to I^*X$, constructed in the slice category $\mathsf{cSet}/_I$, is a trivial fibration. Condition (2) says that the pullback-hom of the generic point $\delta : 1 \to \mathbb{I}$ with I^*f, constructed in the slice category $\mathsf{cSet}/_I$, is a trivial fibration. Thus a map $f : Y \to X$ is an *unbiased* fibration just if its base change I^*f is a δ-*biased* fibration in the slice category $\mathsf{cSet}/_I$. The latter condition can also be reformulated as follows.

Proposition 4.7 *A map $f : Y \to X$ is an unbiased fibration if and only if the canonical map u to the pullback, in the following diagram in* cSet, *is a trivial fibration.*

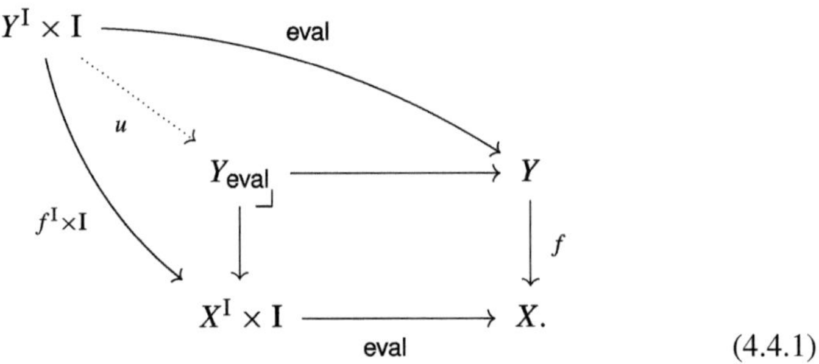

$$(4.4.1)$$

Proof We interpolate another pullback into the rectangle in (4.4.1) to obtain

$$
\begin{array}{ccccc}
Y_{\mathsf{eval}} & \longrightarrow & Y \times I & \longrightarrow & Y \\
\downarrow & & \downarrow & & \downarrow f \\
X^I \times I & \longrightarrow & X \times I & \longrightarrow & X
\end{array}
$$

$$(4.4.2)$$

with the evident maps. The left hand square is therefore a pullback, so we indeed have that

$$Y_{\mathsf{eval}} \cong (X^I \times I) \times_{(X \times I)} (Y \times I) \cong (X^I \times I) \times_X Y$$

and $u = (\delta \Rightarrow f)$. $\square$

As a special case, we have:

Corollary 4.8 *An object X is unbiased fibrant if and only if the canonical map u to the pullback, in the following diagram in* cSet, *is a trivial fibration.*

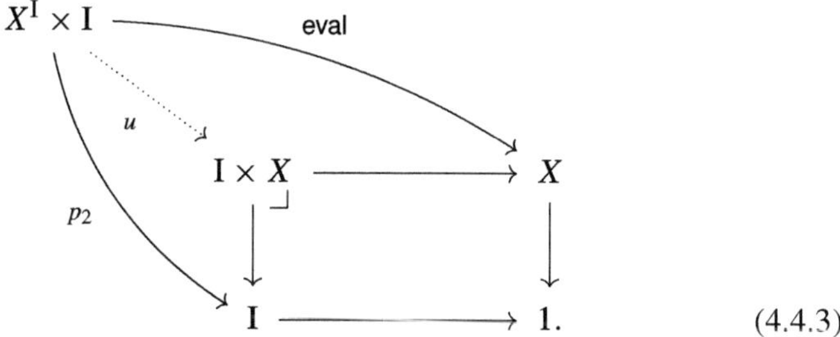

$$(4.4.3)$$

Now we can run the proof of Proposition 4.3 backwards in order to determine a class of generating trivial cofibrations for the unbiased case. Consider pairs of maps $c : C \rightarrowtail Z$ and $i : Z \to I$, where the former is a cofibration and the latter is regarded as an "I-indexing", so that

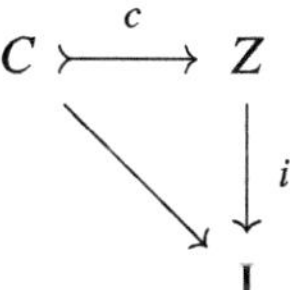

is regarded as an "I-indexed family of cofibrations $c_i : C_i \rightarrowtail Z_i$". We shall use the notation

$$\langle i \rangle := \langle 1_Z, i \rangle : Z \longrightarrow Z \times I, \tag{4.4.4}$$

for the graph of the indexing map $i : Z \to I$. Then write

$$c \otimes_i \delta := [\langle i \rangle, c \times I] : Z +_C (C \times I) \longrightarrow Z \times I,$$

which is easily seen to be well-defined on the indicated pushout below.

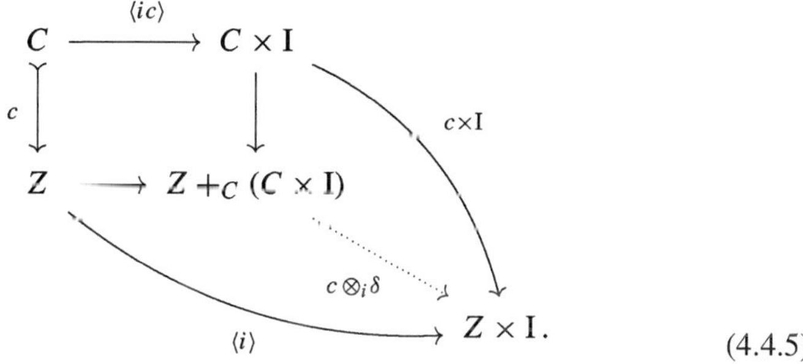

$$(4.4.5)$$

Remark 4.9 The specification (4.4.5) differs from the similar (4.1.2) by using the graph $\langle i \rangle : Z \rightarrowtail Z \times \mathrm{I}$ for the inclusion of Z into the cylinder over Z, rather than one of the two "ends",

$$\langle 1_Z, \delta_\epsilon ! \rangle : Z \cong Z \times 1 \xrightarrow{Z \times \delta_\epsilon} Z \times \mathrm{I} \tag{4.4.6}$$

arising from the endpoint inclusions $\delta_\epsilon : 1 \rightarrow \mathrm{I}$, for $\epsilon = 0, 1$. As an arrow over I, the graph $\langle i \rangle : Z \rightarrow Z \times \mathrm{I}$ also takes the form (4.4.6), namely

$$\langle i \rangle = \langle 1_Z, \delta ! \rangle : Z \rightarrow Z \times \mathrm{I}.$$

If we also regard $c : C \rightarrow Z$ as an arrow over I via $i : Z \rightarrow \mathrm{I}$, and use the generic point $\delta : 1 \rightarrow \mathbb{I}$ over I in place of $\delta_\epsilon : 1 \rightarrow \mathrm{I}$, then (4.4.5) agrees with (4.1.2), up to those changes. Thus the indicated map $c \otimes_i \delta$ in (4.4.5) *is the pushout-product constructed over* I of the generic point δ, which is a cofibration by (C7), with the map c regarded as an I-indexed family of cofibrations via the indexing $i : Z \rightarrow \mathrm{I}$.

Observe that for any map $i : Z \rightarrow \mathrm{I}$, the graph $\langle i \rangle = \langle 1_Z, i \rangle : Z \rightarrow Z \times \mathrm{I}$ is a cofibration, since it is a pullback of the diagonal of I along $i \times \mathrm{I}$. The subobject

$$c \otimes_i \delta \rightarrowtail Z \times \mathrm{I}$$

constructed in (4.4.5) is therefore a cofibration, since it is the join in the lattice $\mathsf{Sub}(Z \times \mathrm{I})$ of the cofibrant subobjects $\langle i \rangle \rightarrowtail Z \times \mathrm{I}$ and $C \times \mathrm{I} \rightarrowtail Z \times \mathrm{I}$, where the latter is the "cylinder over $C \rightarrowtail Z$".

Definition 4.10 The maps of the form $c \otimes_i \delta : Z +_C (C \times \mathrm{I}) \rightarrowtail Z \times \mathrm{I}$ now form the class of *generating unbiased trivial cofibrations*,

$$C \otimes \delta = \{ c \otimes_i \delta : D \rightarrowtail Z \times \mathrm{I} \mid c : C \rightarrowtail Z, i : Z \rightarrow \mathrm{I} \}. \tag{4.4.7}$$

We can then show that the unbiased fibrations are exactly the right class of these maps,

$$(C \otimes \delta)^{\pitchfork} = \mathcal{F}.$$

Proposition 4.11 *A map $f : Y \rightarrow X$ is an unbiased fibration iff for every pair of maps $c : C \rightarrowtail Z$ and $i : Z \rightarrow \mathrm{I}$, where the former is a cofibration, every commutative square of the following form has a diagonal filler, as indicated in the following.*

$$
\begin{array}{ccc}
Z +_C (C \times \mathrm{I}) & \longrightarrow & Y \\
{\scriptstyle c \otimes_i \delta}\Big\downarrow & \nearrow^{\ j} & \Big\downarrow{\scriptstyle f} \\
Z \times \mathrm{I} & \longrightarrow & X.
\end{array}
\tag{4.4.8}
$$

Proof Suppose that for all $c : C \rightarrowtail Z$ and $i : Z \rightarrow I$, we have $(c \otimes_i \delta) \pitchfork f$ in cSet. Pulling f back over I, this is equivalent to the condition $c \otimes \delta \pitchfork I^* f$ in $\mathsf{cSet}_{/\mathrm{I}}$, for all cofibrations $c : C \rightarrowtail Z$ over I, which is equivalent to $c \pitchfork (\delta \Rightarrow I^* f)$ in $\mathsf{cSet}_{/\mathrm{I}}$ for all cofibrations $c : C \rightarrowtail Z$. But this in turn means that $\delta \Rightarrow I^* f$ is a trivial fibration, which by definition means that f is an unbiased fibration. $\qquad\square$

Remark 4.12 A warning is perhaps in order that the collection $C \otimes \delta$ of generating unbiased trivial cofibrations is a proper class; we shall consider the generating subset $\mathsf{BCof} \otimes \delta \subset C \otimes \delta$ in (4.5.2) below.

Note also that the endpoints $\delta_\epsilon : 1 \rightarrow I$, in particular, are of the form $c \otimes_i \delta$ by taking $Z = 1$ and $i = \delta_\epsilon$ and $c = {!} : 0 \rightarrow 1$, so that the case of biased filling is subsumed. Moreover, for any $i : Z \rightarrow I$ the graph $\langle i \rangle : Z \rightarrowtail Z \times I$ is itself of the form $0 \otimes_i \delta$ for the cofibration $0 \rightarrowtail Z$, so the graph of any "I-indexing" map $i : Z \rightarrow I$ is also a trivial cofibration.

The following sanity check will also be needed later.

Proposition 4.13 *Let* $f : F \twoheadrightarrow X$ *be an unbiased fibration in* cSet. *Then for the endpoints* $\delta_0, \delta_1 : 1 \rightarrow I$, *the associated pullback-homs,*

$$\delta_\epsilon \Rightarrow f : F^{\mathrm{I}} \rightarrow X^{\mathrm{I}} \times_X F \qquad (\epsilon = 0, 1) \tag{4.4.9}$$

are also trivial fibrations. Thus unbiased fibrations are also δ_ϵ*-biased fibrations, for* $\epsilon = 0, 1$.

Proof This follows from Remark 4.12 and the $\otimes \dashv \Rightarrow$ adjunction, but we give a different proof. Consider the case $X = 1$, the general one $f : F \rightarrow X$ being analogous. Thus let F be an unbiased fibrant object in cSet. So by definition $(I^* F)^\delta : (I^* F)^{\mathbb{I}} \longrightarrow I^* F$ in $\mathsf{cSet}_{/\mathrm{I}}$ is a trivial fibration. Pulling back $\delta : 1 \rightarrow \mathbb{I}$ in $\mathsf{cSet}_{/\mathrm{I}}$ along the base change $\delta_\epsilon : 1 \rightarrow I$ takes it to $\delta_\epsilon : 1 \rightarrow I$ in cSet, by the universal property of the generic point $\delta : 1 \rightarrow \mathbb{I}$; that is $\delta_\epsilon^*(\delta) = \delta_\epsilon : 1 \rightarrow I$. So $(I^* F)^\delta : (I^* F)^{\mathbb{I}} \longrightarrow I^* F$ is taken by δ_ϵ^* to

$$\delta_\epsilon^*\big((I^* F)^\delta\big) = (\delta_\epsilon^* I^* F)^{\delta_\epsilon^* \delta} = F^{\delta_\epsilon} : F^{\mathrm{I}} \longrightarrow F \,,$$

as shown in the following.

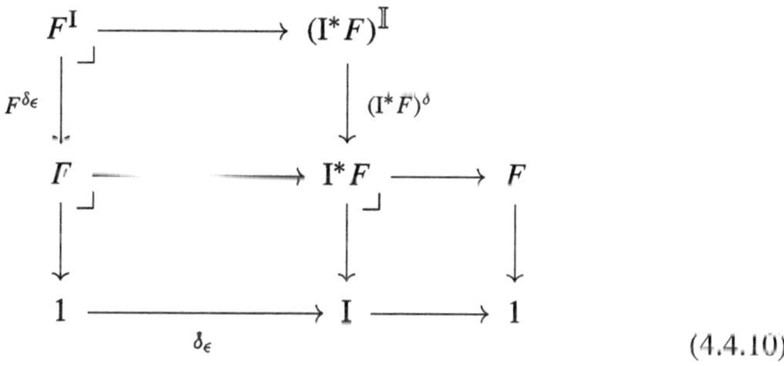

$$\tag{4.4.10}$$

And pullback preserves trivial fibrations.   □

4.5 Unbiased Fibration Structure

As in the biased case, the fibrations can be determined by *uniform* right-lifting against a *small category* of unbiased trivial cofibrations, now consisting of all those $c \otimes_i \delta$ in $C \otimes \delta$ for which $c : C \rightarrowtail I^n$ is basic, i.e. has representable codomain. Call these maps the *basic unbiased trivial cofibrations*, and let

$$\mathsf{BCof} \otimes \delta = \{ c \otimes_i \delta : B \rightarrowtail I^{n+1} \mid c : C \rightarrowtail I^n, \ i : I^n \rightarrow I, \ n \geq 0 \}, \tag{4.5.1}$$

where the pushout-product $c \otimes_i \delta$ now has the form

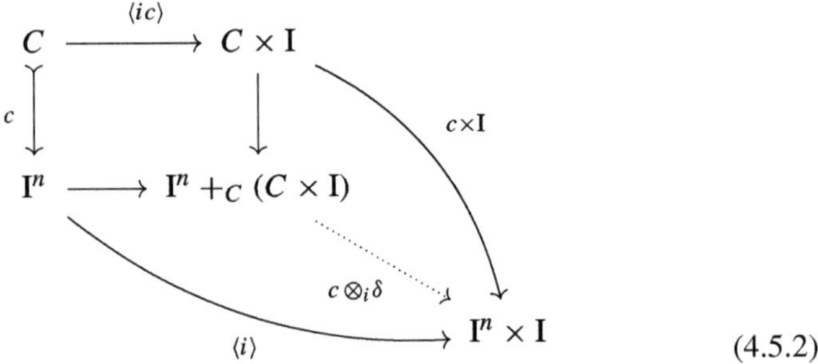

$$\tag{4.5.2}$$

for a basic cofibration $c : C \rightarrowtail I^n$ and an indexing map $i : I^n \rightarrow I$, and with domain $B = \left(I^n +_C (C \times I) \right)$. These subobjects $B \rightarrowtail I^{n+1}$ can again be seen geometrically as "generalized open box inclusions", but now the floor and lid of the open box are generalized to the graph of an arbitrary map $i : I^n \rightarrow I$.

For any map $f : Y \rightarrow X$ a uniform, unbiased fibration structure on f is then a choice of diagonal fillers $j(c, i, x, y)$,

$$\tag{4.5.3}$$

for each basic trivial cofibration $c \otimes_i \delta : B \rightarrowtail I^{n+1}$, which is *uniform in I^n* in the following sense: Given any cubical map $u : I^m \rightarrow I^n$, the pullback $u^*c : u^*C \rightarrowtail I^m$ and the reindexing $iu : I^m \rightarrow I^n \rightarrow I$ determine another basic trivial cofibration $u^*c \otimes_{iu} \delta : B' = (I^m +_{u^*C} (u^*C \times I)) \rightarrowtail I^{m+1}$, which fits into a commutative diagram of the form

$$
\begin{array}{ccccc}
B' & \xrightarrow{\ (u\times\mathrm{I})'\ } & B & \xrightarrow{\ x\ } & X \\
{\scriptstyle u^*c\,\otimes_{iu}\delta}\big\downarrow & \lrcorner & {\scriptstyle c\,\otimes_i\delta}\big\downarrow & {}^{\,j(c,i,x,y)} & \big\downarrow{\scriptstyle f} \\
\mathrm{I}^m\times\mathrm{I} & \xrightarrow[\ u\times\mathrm{I}\ ]{} & \mathrm{I}^n\times\mathrm{I} & \xrightarrow[\ y\ ]{} & Y.
\end{array}
\tag{4.5.4}
$$

For the outer rectangle in (4.5.4) there is a chosen diagonal filler

$$
j(u^*c, iu, x(u\times\mathrm{I})', y(u\times\mathrm{I})) : \mathrm{I}^m\times\mathrm{I}\to X,
$$

and for this map we require that

$$
j(u^*c, iu, x(u\times\mathrm{I})', y(u\times\mathrm{I})) = j(c, i, x, y)\circ(u\times\mathrm{I}).
\tag{4.5.5}
$$

Definition 4.14 A *uniform, unbiased fibration structure* on a map

$$
f : Y\to X
$$

is a choice of fillers $j(c, i, x, y)$ as in (4.5.3) satisfying (4.5.5) for all cubical maps $u : \mathrm{I}^m\to\overline{\mathrm{I}}^n$.

In these terms, we have the following analogue of Corollary 4.5.

Proposition 4.15 *For any object X in* cSet *the following are equivalent:*

1. *X is an unbiased fibrant object in the sense of Definition 4.6: the canonical map $\delta\Rightarrow X : X^{\mathrm{I}}\times\mathrm{I}\to X\times\mathrm{I}$ is a trivial fibration.*
2. *X has the right lifting property with respect to all generating unbiased trivial cofibrations,*

$$
(C\otimes\delta)\ \pitchfork\ X.
$$

3. *X has a uniform, unbiased fibration structure in the sense of Definition 4.14.*

Proof The equivalence between (1) and (2) is Proposition 4.11. So assume (1). Then in cSet$/_{\mathbb{1}}$, the evaluation at $\delta : 1\to\mathbb{I}$,

$$
(\mathrm{I}^*X)^\delta : (\mathrm{I}^*X)^{\mathbb{I}}\longrightarrow X
$$

is a trivial fibration. By Proposition 3.9 it therefore has a uniform filling structure with respect to all basic cofibrations $c : C\rightarrowtail\mathrm{I}^n$ over I. Transposing by the $\otimes\dashv\Rightarrow$ adjunction and unwinding then gives exactly a uniform fibration structure on X. $\qquad\square$

A statement analogous to the foregoing also holds for maps $f : Y \to X$ in place of objects X. Indeed, as before, we have the following sharper formulation.

Corollary 4.16 *Uniform, unbiased fibration structures on a map* $f : Y \to X$ *correspond uniquely to relative* $+$-*algebra structures on the map* $(\delta \Rightarrow f)$ *(cf. Definition 4.6),*

$$(\delta \Rightarrow f) : Y^I \times \mathrm{I} \longrightarrow (X^I \times \mathrm{I}) \times_{(X \times \mathrm{I})} (Y \times \mathrm{I}).$$

4.6 Factorization

Definition 4.17 Summarizing the foregoing definitions, we have the following classes of maps:

- The *generating unbiased trivial cofibrations* were determined in (4.4.7) as the class

$$C \otimes \delta = \{ c \otimes_i \delta : D \rightarrowtail Z \times \mathrm{I} \mid c : C \rightarrowtail Z, \ i : Z \to \mathrm{I} \}, \tag{4.6.1}$$

where $D = \big(Z +_C (C \times \mathrm{I}) \big)$ and the pushout-product $c \otimes_i \delta$ has the form

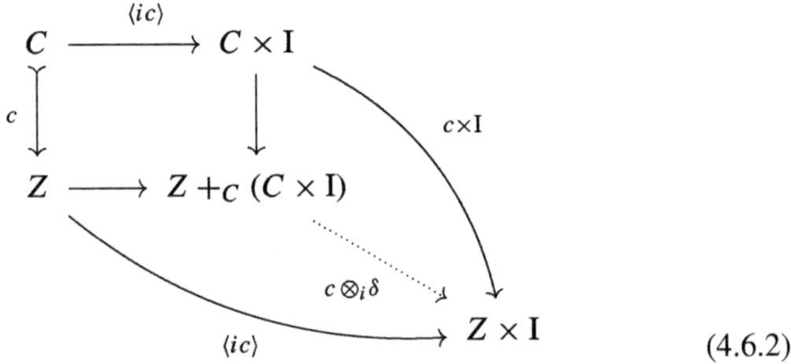

$$\tag{4.6.2}$$

for any cofibration $c : C \rightarrowtail Z$ and indexing map $i : Z \to \mathrm{I}$.
- The class $\mathcal{F}$ of *unbiased fibrations*, which can be characterized as the right-lifting class of the generating unbiased trivial cofibrations,

$$(C \otimes \delta)^{\pitchfork} = \mathcal{F}.$$

- The class of *unbiased trivial cofibrations* is then defined to be left-lifting class of the fibrations,

$$\mathsf{TCof} = {}^{\pitchfork}\mathcal{F}.$$

It follows that the classes TCof and $\mathcal{F}$ are closed under retracts and are mutually weakly orthogonal, $\mathsf{TCof} \pitchfork \mathcal{F}$. Thus in order to have a weak factorization system $(\mathsf{TCof}, \mathcal{F})$ it just remains to show the following.

Lemma 4.18 *Every map* $f : X \to Y$ *in* cSet *can be factored as* $f = p \circ i$,

$$(4.6.3)$$

with $i : X \rightarrowtail X'$ *an unbiased trivial cofibration and* $p : X' \twoheadrightarrow Y$ *an unbiased fibration.*

Proof We can use a standard argument (the "algebraic small object argument", cf. [36, 64]), which can be further simplified using the fact that the codomains of the basic trivial cofibrations $c \otimes_i \delta : B \rightarrowtail I^{n+1}$ from (4.5.1) are not just representable, but *tiny* in the sense of Proposition 2.3, and the domains are not merely "small", but *finitely presented*. Note in particular that the collection $\mathsf{BCof} \otimes \delta$ is a set. The reader is referred to [6] for details in a similar case. $\qquad\square$

Remark 4.19 The proof in ibid. actually produces a stronger result than we need, namely an *algebraic* weak factorization system. This follows from the small generating *category* $\mathsf{BCof} \otimes \delta$ of basic unbiased trivial cofibrations (and pullback squares of the form on the left in (4.5.4)). The relationship between this stronger condition and the *classifying types* used in Chap. 7 is studied in [69], which also gives an even more "constructive" proof of the factorization Lemma 4.18, not requiring quotients, exactness, or impredicativity. With this modification, the present approach can also be used in a *quasi*-topos, as occurs in e.g. realizability and sheaves.

Proposition 4.20 *There is a weak factorization system on the category* cSet *in which the right maps are the unbiased fibrations and the left maps are the unbiased trivial cofibrations, both as specified in Definition 4.17. This will be called the* (unbiased) fibration weak factorization system.

Hereafter, unless otherwise stated, all fibrations in cSet are assumed to be unbiased.

Chapter 5
The Weak Equivalences

Our approach to proving that the classes C and $\mathcal{F}$ of cofibrations and fibrations, as defined in Chaps. 3 and 4, determine a model structure on cSet will be to first identify a *premodel structure* in the sense of [15], and then turn to the question of the 3-for-2 property for the resulting weak equivalences.

Definition 5.1 (Weak Equivalence) A map $f : X \to Y$ in cSet is a *weak equivalence* if it can be factored as $f = g \circ h$,

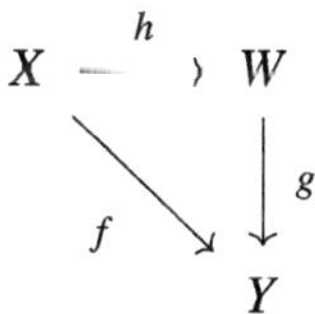

with $h \pitchfork \mathcal{F}$ and $C \pitchfork g$. Accordingly, let

$$\mathcal{W} = \mathsf{TFib} \cup \mathsf{TCof}$$

$$= \{f : X \to Y \mid f = g \circ h \text{ for some } g \in \mathsf{TFib} \text{ and } h \in \mathsf{TCof}\}$$

be the class of weak equivalences.

Observe first that every trivial fibration $f \in \mathsf{TFib} = C^{\pitchfork}$ is indeed a fibration, because the generating trivial cofibrations $c \otimes_i \delta$ are cofibrations. Moreover, every trivial fibration $f : X \to Y$ is also a weak equivalence $f = f \circ 1_X$, since the identity map 1_X is (trivially) a trivial cofibration $\mathsf{TCof} = {}^{\pitchfork}\mathcal{F}$. Thus we have

$$\mathsf{TFib} \subseteq (\mathcal{F} \cap \mathcal{W}).$$

© The Author(s), under exclusive license to Springer Nature Switzerland AG 2026
S. Awodey, *Cartesian Cubical Model Categories*, Lecture Notes
in Mathematics 2385, https://doi.org/10.1007/978-3-032-08730-0_5

Similarly, because $\mathsf{TFib} \subseteq \mathcal{F}$, we have $\mathsf{TCof} \subseteq C$. Moreover, since identity maps are also trivial fibrations we have $\mathsf{TCof} \subseteq \mathsf{TFib} \circ \mathsf{TCof} = \mathcal{W}$. Thus we also have

$$\mathsf{TCof} \subseteq (C \cap \mathcal{W}).$$

Lemma 5.2 $(C \cap \mathcal{W}) \subseteq \mathsf{TCof}$.

Proof Let $c : A \rightarrowtail B$ be a cofibration with a factorization

$$c = tf \circ tc : A \to W \to B$$

where $tc \in \mathsf{TCof}$ and $tf \in \mathsf{TFib}$. Let $f : X \twoheadrightarrow Y$ be a fibration and consider a commutative diagram,

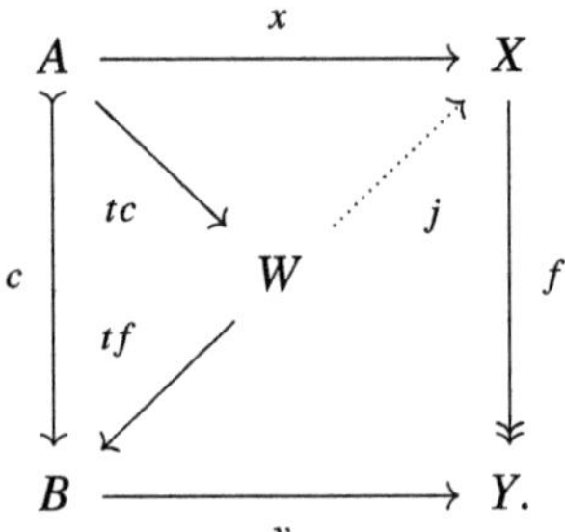

Inserting the factorization of c, from $tc \pitchfork f$ we obtain $j : W \to X$ as indicated, with $j \circ tc = x$ and $f \circ j = y \circ tf$.

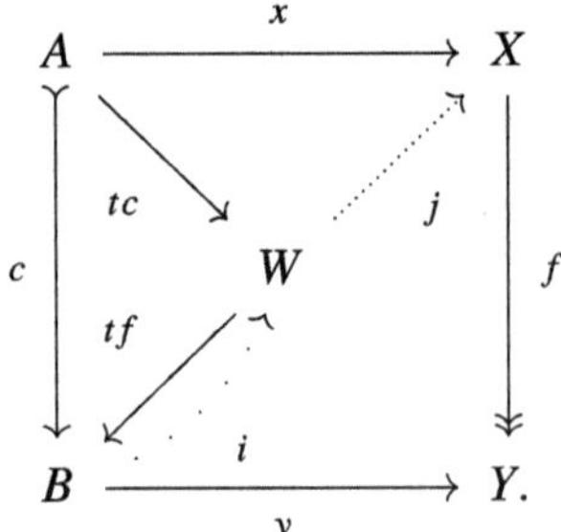

Moreover, since $c \pitchfork tf$ there is an $i : B \to W$ as indicated, with $i \circ c = tc$ and $tf \circ i = 1_B$.

Let $k = j \circ i$. Then $k \circ c = j \circ i \circ c = j \circ t c = x$, and $f \circ k = f \circ j \circ i = y \circ t \circ f \circ i = y$.

$$\square$$

The proof of the following is exactly dual.

Lemma 5.3 $(\mathcal{F} \cap \mathcal{W}) \subseteq \mathsf{TFib}.$

Proposition 5.4 *The three classes of maps* $C, \mathcal{W}, \mathcal{F}$ *in* cSet *constitute a* premodel structure *in the sense of [15]. In particular, we have*

$$\mathcal{F} \cap \mathcal{W} = \mathsf{TFib},$$

$$C \cap \mathcal{W} = \mathsf{TCof},$$

and therefore two interlocking weak factorization systems:

$$(C, \, \mathcal{W} \cap \mathcal{F}) \, , \quad (C \cap \mathcal{W}, \, \mathcal{F}).$$

It now "only" remains to show that the weak equivalences $\mathcal{W}$ satisfy the 3-for-2 axiom of Definition 1.1 in order to verify that $(C, \mathcal{W}, \mathcal{F})$ is a model structure. Perhaps surprisingly, this will occupy the remainder of these lecture notes! We shall follow roughly the approach of [47]: the weak equivalences between fibrant objects are shown to be the usual *homotopy equivalences*, which evidently satisfy 3-for-2. So we reduce to this case using the fact that K^X is fibrant whenever K is. It suffices, namely, to show that the weak equivalences are those maps $w : X \to Y$ that induce homotopy equivalences $K^w : K^Y \simeq K^X$ for fibrant K. Such maps are termed *weak homotopy equivalences* (Definition 5.12), and our task will therefore be to show that a map is a weak equivalence if and only if it is a weak homotopy equivalence.

5.1 Homotopy Equivalence

The following definition is the standard one from homotopy theory, using the cubical interval $\delta_0, \delta_1 : 1 \rightrightarrows \mathrm{I}$.

Definition 5.5 (Homotopy) A *homotopy* $\vartheta : f \sim g$ between maps $f, g : X \rightrightarrows Y$ is a map,

$$\vartheta : \mathrm{I} \times X \longrightarrow Y,$$

such that $\vartheta \circ \iota_0 = f$ and $\vartheta \circ \iota_1 = g$,

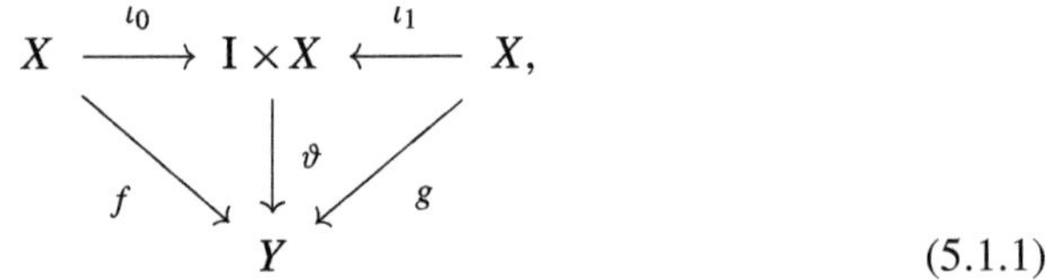

$$(5.1.1)$$

where ι_0, ι_1 are the canonical inclusions into the ends of the cylinder,

$$\iota_\epsilon : X \cong 1 \times X \xrightarrow{\;\;\delta_\epsilon \times X\;\;} I \times X , \qquad \epsilon = 0, 1.$$

Note that each of the inclusions $\iota_\epsilon : X \rightarrowtail I \times X$ is a cofibration, as is their join $X + X \rightarrowtail I \times X$, by Remark 4.1.

Proposition 5.6 *The relation of homotopy $f \sim g$ between maps $f, g : X \rightrightarrows Y$ is preserved by pre- and post-composition. If Y is fibrant, then $f \sim g$ is an equivalence relation.*

Proof Inspecting (5.1.1), preservation of $f \sim g$ under post-composing with any $h : Y \rightarrow Z$ is obvious: we have $h \circ \vartheta : h \circ f \sim h \circ g$. Now observe that a homotopy $f \overset{\vartheta}{\sim} g : X \times I \rightarrow Y$ determines a (unique) path $\tilde{\vartheta} : I \rightarrow Y^X$ in the function space, with endpoints $\vartheta_0 = \vartheta \circ \delta_0 = \tilde{f} : 1 \rightarrow Y^X$ and $\vartheta_1 = \vartheta \circ \delta_1 = \tilde{g}$. Precomposing maps $f, g : X \rightrightarrows Y$ with any $e : W \rightarrow X$ is induced by post-composing $\tilde{f}, \tilde{g} : 1 \rightarrow Y^X$ with the map $Y^e : Y^X \rightarrow Y^W$, which then also takes the path $\tilde{\vartheta} : I \rightarrow Y^X$ to a path $\tilde{\varphi} = Y^e \circ \tilde{\vartheta} : I \rightarrow Y^W$ corresponding to a (unique) homotopy $\varphi : f \circ e \sim g \circ e$.

Now note that Y^X is fibrant if Y is fibrant, since the generating trivial cofibrations $c \times_i \delta$ are preserved by the functor $X \times (-)$. So we can use "box filling" in Y^X to verify the claimed equivalence relation.

- Reflexivity $f \sim f$ is witnessed by the homotopy $\rho : I \rightarrow 1 \xrightarrow{f} Y^X$.
- For symmetry $f \sim g \Rightarrow g \sim f$ take $\vartheta : I \rightarrow Y^X$ with $\vartheta_0 = f$ and $\vartheta_1 = g$ and we want to build $\vartheta' : I \rightarrow Y^X$ with $\vartheta'_0 = g$ and $\vartheta'_1 = f$. Take an open 2-box in Y^X of the following form.

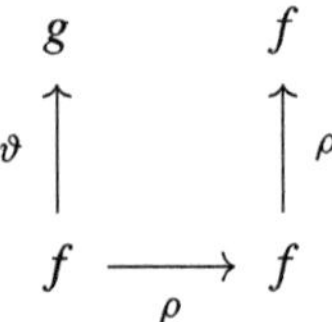

This box is a map $b : I +_1 I +_1 I \rightarrow Y^X$ with the indicated components, and it has a filler $c : I \times I \rightarrow Y^X$, i.e. an extension along the canonical map $I +_1 I +_1 I \rightarrowtail I \times I$, which is a trivial cofibration of the form $\partial I \otimes \delta_0$. Let $t : I \rightarrow I \times I$ be the top face of the 2-cube (the bipointed map $\{0, x_1, x_2, 1\} \to \{0, x, 1\}$ that is constantly 1). We can set $\vartheta' = c \circ t : I \to Y^X$ to get a homotopy $\vartheta' : I \to Y^X$ with $\vartheta'_0 = g$ and $\vartheta'_1 = f$ as required.

- For transitivity, $f \overset{\vartheta}{\sim} g$, $g \overset{\varphi}{\sim} h \Rightarrow f \sim h$, an analogous construction will fill the open box:

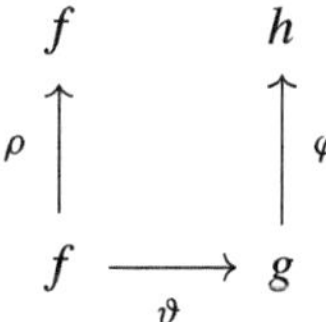

$$\square$$

We then have the usual definition of homotopy equivalence:

Definition 5.7 (Homotopy Equivalence)   A *homotopy equivalence* is a map $f : X \to Y$ together with a map $g : Y \to X$ and homotopies $\vartheta : 1_X \sim g \circ f$ and $\varphi : 1_Y \sim f \circ g$. We call g a *quasi-inverse* of f.

Since the quasi-inverses are also homotopy equivalences, these maps will satisfy the 3-for-2 condition just if they compose, which they do when restricted to fibrant objects:

Lemma 5.8   *The homotopy equivalences between fibrant objects satisfy 3-for-2.*

Proof   Composing two homotopy equivalences $f_1 : X \to Y$ and $f_2 : Y \to Z$ results in a composable pair of homotopies $\vartheta_1 : 1_X \sim g_1 \circ f_1$ and $g_1 \vartheta_2 f_1 : g_1 \circ f_1 \sim g_1 \circ g_2 \circ f_2 \circ f_1$. Now use fibrancy of X to compose them pointwise. The case of the other composites at Z is similar. $\square$

Lemma 5.9   *A fibration that is a weak equivalence is a homotopy equivalence.*

Proof   Any trivial fibration $f : X \twoheadrightarrow Y$ has a section $s : Y \to X$ by Corollary 3.10. Consider the following lifting problem:

$$
\begin{array}{ccc}
X + X & \xrightarrow{\;[sf,1]\;} & X \\
{\scriptstyle [\iota_0, \iota_1]}\downarrow & & \downarrow{\scriptstyle f} \\
I \times X & \xrightarrow[\;f\pi_2\;]{} & Y
\end{array}
$$

Since the map on the left is a cofibration, a diagonal filler provides a homotopy $\vartheta : sf \sim 1_X$. Thus f is a homotopy equivalence. $\square$

For the further comparison of the weak equivalences with the homotopy equivalences we need the following.

5.2   Weak Homotopy Equivalence

Definition 5.10 (Connected Components)   The functor

$$\pi_0 : \mathsf{cSet} \to \mathsf{Set}$$

is defined on a cubical set X as the coequalizer

$$X_1 \rightrightarrows X_0 \to \pi_0 X \,,$$

where the two parallel arrows are the maps $X_{\delta_0}, X_{\delta_1} : X_1 \rightrightarrows X_0$ for the endpoints $\delta_0, \delta_1 : 1 \rightrightarrows \mathrm{I}$. If K is fibrant, then by the foregoing Proposition 5.6, for any X we have

$$\pi_0(K^X) = \mathrm{Hom}(X, K)/\sim \,.$$

That is, $\pi_0(K^X)$ is the set $[X, K]$ of homotopy equivalence classes of maps $X \to K$.

Remark 5.11   One can show that in fact $\pi_0 X = \varinjlim X_n$ where the colimit is taken over *all* objects $[n]$ in the index category $\square^{\mathrm{op}} = \overrightarrow{\mathbb{B}}$, rather than just the "last" two $[1] \rightrightarrows [0]$. Since the category $\mathbb{B}$ of finite strictly bipointed sets is sifted, the functor $\pi_0 : \mathsf{cSet} \to \mathsf{Set}$ preserves finite products.

Definition 5.12 (Weak Homotopy Equivalence)   A map $f : X \to Y$ is called a *weak homotopy equivalence* if for every fibrant object K, the canonical map $K^f :$ $K^Y \to K^X$ is bijective on connected components,

$$\pi_0(K^f) : \pi_0(K^Y) \cong \pi_0(K^X) \,.$$

Lemma 5.13   *Every homotopy equivalence is a weak homotopy equivalence.*

Proof   Let $f : X \to Y$ be a homotopy equivalence. Then $K^f : K^Y \to K^X$ is also a homotopy equivalence for any K, since homotopy respects (pre- and) post-composition by all maps. If K is fibrant, then so is K^X and π_0 is well defined on homotopy classes of maps, by Proposition 5.6. It clearly takes homotopy equivalences to isomorphisms of sets, since it identifies homotopic maps. $\square$

Lemma 5.14   *The weak homotopy equivalences also satisfy the 3-for-2 condition.*

Proof   This follows by applying the Set-valued functors $\pi_0(K^{(-)})$, for all fibrant objects K, and the corresponding fact about bijections of sets. $\square$

In virtue of Lemma 5.14 it now suffices to show that a map is a weak equivalence if and only if it is a weak homotopy equivalence. The following characterization will be useful.

Lemma 5.15   *A map $f : X \to Y$ is a weak homotopy equivalence just if it satisfies the following two conditions.*

1. *For every fibrant object K and every map $x : X \to K$ there is a map $y : Y \to K$ such that $y \circ f \sim x$,*

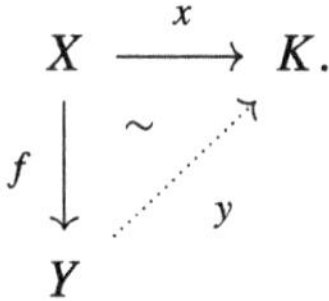

We say that x "extends along f up to homotopy".

2. *For every fibrant object K and maps $y, y' : Y \to K$ such that $yf \sim y'f$, there is a homotopy $y \sim y'$,*

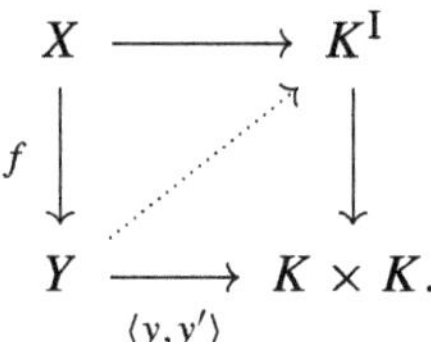

Proof Condition (1) says exactly that the internal precomposition map $K^f : K^Y \to K^X$ is surjective under connected components π_0, while (2) says just that it is injective under π_0. □

Lemma 5.16 *Any weak equivalence is a weak homotopy equivalence.*

Proof By Lemmas 5.9 and 5.13, a trivial fibration is also a weak homotopy equivalence. So it suffices to consider the trivial cofibrations, since weak homotopy equivalences are closed under composition, by Lemma 5.14. Thus let $f : X \rightarrowtail Y$ be a trivial cofibration, and apply Lemma 5.15: condition (1) is immediate, and (2) follows because $K^I \twoheadrightarrow K \times K$ is a fibration when K is fibrant, since $\partial : 1 + 1 \rightarrowtail I$ is a cofibration (by Remark 4.1). □

Our goal is now to show the converse of Lemma 5.16(2), that a weak homotopy equivalence is a weak equivalence. We shall first restrict attention to maps $f : X \to K$ with a fibrant codomain K. By factoring such maps, we can split into the cases of a fibration and a cofibration.

Lemma 5.17 *If K is fibrant, then any fibration $f : X \twoheadrightarrow K$ that is a homotopy equivalence is a weak equivalence.*

Proof This is a standard argument, which we just sketch. It suffices to show that any diagram of the form

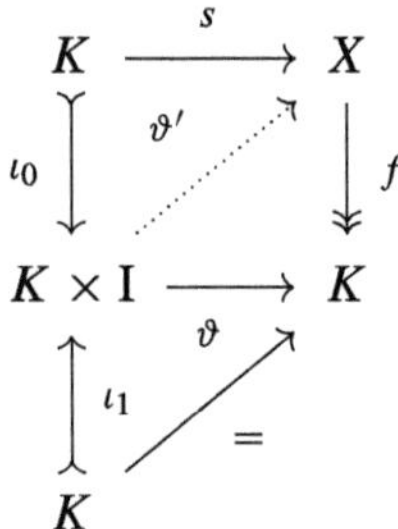

$$(5.2.1)$$

with $c : C \rightarrowtail X$ a cofibration, has a diagonal filler, for then f is a trivial fibration. Since f is a homotopy equivalence, it has a quasi-inverse $s : K \to X$ with $\vartheta : fs \sim 1_K$, which we claim can be corrected to a section $s' : K \to X$. Indeed, consider

where ϑ' results from $\iota_0 \pitchfork f$. Let $s' = \vartheta'\iota_1$, so that $\vartheta' : s \sim s'$ and $fs' = 1_K$.

Thus we can assume that $s = s' : K \to X$ is a section, which fills the diagram (5.2.1) up to a homotopy in the upper triangle.

Now we can correct $s : K \to X$ to a homotopic $t : K \to X$ over f by using the homotopy $\varphi : sc \sim x$ to get a map $\varphi : C \to X^I$ over f. Since f is a fibration, the projections $p_0, p_1 : X^I \to X$ over f are trivial fibrations, and so there is a lift $\varphi' : K \to X^I$ for which $t := p_1\varphi'$ has $tc = x$ and $ft = 1_K$, and so is a filler for (5.2.1).   □

Lemma 5.18 *If K is fibrant, then any fibration $f : X \twoheadrightarrow K$ that is a weak homotopy equivalence is a weak equivalence.*

Proof Since K is fibrant, so is X, and since f is a weak homotopy equivalence, by Lemma 5.15(1) there is then a map $s : K \to X$ and a homotopy $\theta : sf \sim 1_X$. Postcomposing with f gives a homotopy $f\vartheta : fsf \sim f$, forming the outer

commutative square in

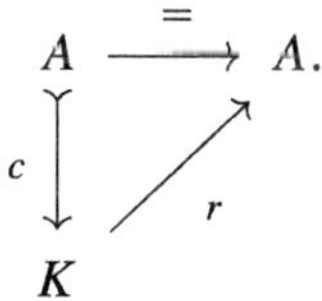

By Lemma 5.15(2) there is a diagonal filler $\varphi : fs \sim 1_K$, and so f is a homotopy equivalence. Now apply Lemma 5.17.   □

We now have the following.

Proposition 5.19 *If A and K are both fibrant, then for any cofibration $c : A \rightarrowtail K$ the following are equivalent.*

1. *$c : A \rightarrowtail K$ is a weak equivalence.*
2. *$c : A \rightarrowtail K$ is a homotopy equivalence.*
3. *$c : A \rightarrowtail K$ is a weak homotopy equivalence.*

Proof Suppose (1), so $c : A \rightarrowtail K$ is a trivial cofibration. Then since A is fibrant, it has a retraction $r : K \to A$.

$$
\begin{array}{ccc}
A & \xrightarrow{\ =\ } & A. \\
c \downarrow & \nearrow{\scriptstyle r} & \\
K & &
\end{array}
$$

Since K is fibrant, $K^I \to K \times K$ is a fibration. So the following has a diagonal filler, which is a homotopy $1_K \sim cr$.

$$
\begin{array}{ccccc}
A & \xrightarrow{\ c\ } & K & \xrightarrow{\ K^!\ } & K^I \\
c \downarrow & & & \nearrow{\scriptstyle \vartheta} & \downarrow {\scriptstyle \langle K^{d_0}, K^{d_1}\rangle} \\
K & & \xrightarrow[\ \langle 1_K, cr\rangle\ ]{} & & K \times K
\end{array}
$$

$(2) \Rightarrow (3)$ is Lemma 5.13.

Suppose (3), that $c : A \rightarrowtail K$ is a weak homotopy equivalence. Factor $c = f \circ tc$ with a trivial cofibration $tc : A \rightarrowtail C$ followed by a fibration $f : C \twoheadrightarrow K$. By parts (1) and (2), $tc : A \rightarrowtail C$ is then a weak homotopy equivalence. By 3-for-2 for weak homotopy equivalences, Lemma 5.14, $f : C \twoheadrightarrow K$ is then also a weak homotopy equivalence. By Lemma 5.18, $f : C \twoheadrightarrow K$ is then a weak equivalence.   □

Proposition 5.20 *For fibrations* $f : X \twoheadrightarrow K$ *with fibrant codomain* K, *all three concepts coincide: weak equivalences, weak homotopy equivalences, and homotopy equivalences.*

Proof Let K be fibrant and suppose that $f : X \twoheadrightarrow K$ is a weak homotopy equivalence. Then it is a weak equivalence by Lemma 5.18. By Lemma 5.16 any fibration weak equivalence is a homotopy equivalence, and by Lemma 5.13 any homotopy equivalence is a weak homotopy equivalence. $\square$

Corollary 5.21 *For all maps* $f : X \to Y$ *between fibrant objects* X *and* Y, *all three concepts coincide: weak equivalence, weak homotopy equivalence, and homotopy equivalence.*

Proof Let X and Y be fibrant and factor $f = tf \circ tc$ with a trivial cofibration $tc : X \rightarrowtail F$ followed by a trivial fibration $tf : F \twoheadrightarrow Y$. Then by Proposition 5.19, $tc : X \rightarrowtail F$ is a homotopy equivalence, and by Proposition 5.20 so is $tf : F \twoheadrightarrow Y$, thus $f = tf \circ tc$ is a homotopy equivalence. Again by Lemma 5.13, any homotopy equivalence is a weak homotopy equivalence, and weak homotopy equivalence between fibrant objects is clearly a weak equivalence, by factoring and using the foregoing Propositions 5.19 and 5.20. $\square$

Lemma 5.22 *If* K *is fibrant, then any cofibration* $c : A \rightarrowtail K$ *that is a weak homotopy equivalence is a weak equivalence.*

Proof Let $c : A \rightarrowtail K$ be a cofibration weak homotopy equivalence and factor it into a trivial cofibration $i : A \rightarrowtail Z$ followed by a fibration $p : Z \twoheadrightarrow K$. By Lemma 5.15, any trivial cofibration is clearly a weak homotopy equivalence. So both c and i are weak homotopy equivalences, and therefore so is p by 3-for-2 for weak homotopy equivalences. Since K is fibrant, p is a trivial fibration by Lemma 5.18, and thus c is a weak equivalence. $\square$

It now follows that a weak homotopy equivalence $f : X \to K$ with a fibrant codomain is a weak equivalence. To eliminate the condition on the codomain we use the following lemma due to [27, 2.4.30].

Lemma 5.23 *A cofibration* $c : A \rightarrowtail B$ *weak homotopy equivalence lifts against any fibration* $f : Y \twoheadrightarrow K$ *with fibrant codomain.*

Proof Let $c : A \rightarrowtail B$ be a cofibration weak homotopy equivalence and $f : Y \twoheadrightarrow K$ a fibration with fibrant codomain K, and consider a lifting problem

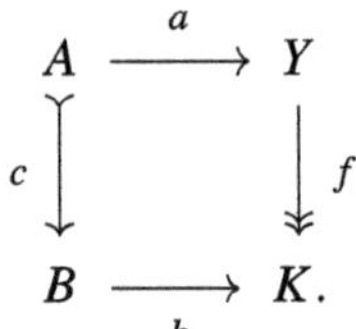

$$
\begin{array}{ccc}
A & \xrightarrow{\ a\ } & Y \\
{\scriptstyle c}\downarrow & & \downarrow{\scriptstyle f} \\
B & \xrightarrow{\ b\ } & K.
\end{array}
$$

Let $\eta : B \rightarrowtail B'$ be a fibrant replacement of B, since K is fibrant, b extends along η to give $b' : B' \to K$ as shown below.

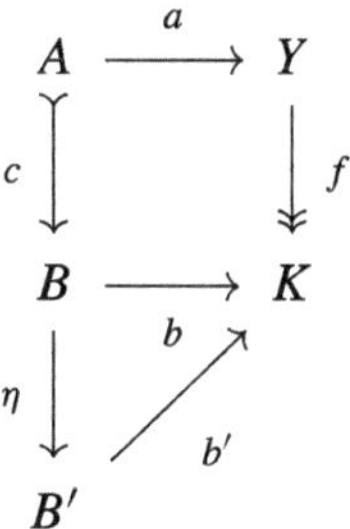

Since η is a trivial cofibration, it is a weak homotopy equivalence. So the composite ηc is also a weak homotopy equivalence. But since B' is fibrant, ηc is then a trivial cofibration by Lemma 5.22. Thus there is a lift $j : B' \to Y$, and therefore also one $k = j\eta : B \to Y$.   □

To complete the proof that a weak homotopy equivalence is a weak equivalence, we shall make use of the following *fibration extension property*, the proof of which is deferred to Chap. 9.

Definition 5.24 (Fibration Extension Property)   For any fibration $f : Y \twoheadrightarrow X$ and any trivial cofibration $\eta : X \to X'$, there is a fibration $f' : Y' \twoheadrightarrow X'$ that pulls back to f along η, as shown below.

$$
\begin{array}{ccc}
Y & \longrightarrow & Y' \\
{\scriptstyle f}\downarrow & & \downarrow{\scriptstyle f'} \\
X & \rightarrowtail & X'
\end{array}
\qquad (5.2.2)
$$

Lemma 5.25   *Assuming the fibration extension property, a cofibration that lifts against every fibration $f : Y \twoheadrightarrow K$ with fibrant codomain is a weak equivalence.*

Proof   Let $c : A \rightarrowtail B$ be a cofibration and consider a lifting problem against an arbitrary fibration $f : Y \twoheadrightarrow X$,

$$
\begin{array}{ccc}
A & \xrightarrow{\;a\;} & Y \\
{\scriptstyle \iota}\downarrow & & \downarrow{\scriptstyle f} \\
B & \xrightarrow{\;b\;} & X.
\end{array}
\qquad (5.2.3)
$$

Let $\eta : X \to X'$ be a fibrant replacement, so η is a trivial cofibration and X' is fibrant. By the fibration extension property of Definition 5.24, there is a fibration

$f' : Y' \twoheadrightarrow X'$ such that f is a pullback of f' along η. So we can extend diagram (5.2.3) to obtain the following, in which the righthand square is a pullback.

$$
\begin{array}{ccccc}
A & \xrightarrow{\ a\ } & Y & \xrightarrow{\ y\ } & Y' \\
\downarrow{\scriptstyle c} & & \downarrow{\scriptstyle f} & & \downarrow{\scriptstyle f'} \\
B & \xrightarrow[\ b\ ]{} & X & \xrightarrow[\ \eta\ ]{} & X'.
\end{array}
\tag{5.2.4}
$$

By assumption, there is a lift $j' : B \to Y'$ with $f'j' = \eta b$ and $j'c = yb$. Therefore, since f is a pullback, there is a map $j : B \to Y$ with $fj = b$ and $yj = j'$.

$$
\begin{array}{ccccc}
A & \xrightarrow{\ a\ } & Y & \xrightarrow{\ y\ } & Y' \\
\downarrow{\scriptstyle c} & \nearrow{\scriptstyle j} & \downarrow & & \downarrow{\scriptstyle f'} \\
B & \xrightarrow[\ b\ ]{} & X & \xrightarrow[\ \eta\ ]{} & X'.
\end{array}
\tag{5.2.5}
$$

Thus $yjc = j'c = ya$. But as a trivial cofibration, η is monic, and as a pullback of η, y is also monic. So $jc = a$. $\square$

Corollary 5.26 *Assuming the fibration extension property,*

1. *a cofibration $c : A \rightarrowtail B$ weak homotopy equivalence is a weak equivalence,*
2. *a fibration $f : Y \twoheadrightarrow X$ weak homotopy equivalence is a weak equivalence.*

Proof (1) follows immediately by combining the previous Lemmas 5.23 and 5.25.

For (2), factor $f : Y \twoheadrightarrow X$ into a cofibration $i : Y \rightarrowtail Z$ followed by a trivial fibration $p : Z \twoheadrightarrow X$. Then f is itself a trivial fibration if $i \pitchfork f$, for then it is a retract of p. Since p is a trivial fibration, it is a weak homotopy equivalence by Lemma 5.16. Since f is also a weak homotopy equivalence, so is i by Lemma 5.14. Thus i is a trivial cofibration by (1). Since f is a fibration, $i \pitchfork f$ as required. $\square$

We have now shown:

Proposition 5.27 *Assuming the fibration extension property, a map $f : X \to Y$ is a weak homotopy equivalence if and only if it is a weak equivalence. Under the same assumption, the weak equivalences $\mathcal{W}$ then satisfy the 3-for-2 condition.*

The results of this Chapter are summarized in the following.

Theorem 5.28 *If the fibration weak factorization system of Definition 4.17 satisfies the fibration extension property of Definition 5.24, then the weak equivalences $\mathcal{W}$ have the 3-for-2 property. Under the same assumption, the classes $(C, \mathcal{W}, \mathcal{F})$ of Proposition 5.4 then form a Quillen model structure. The weak equivalences $\mathcal{W}$ are the weak homotopy equivalences: those maps $f : X \to Y$ for which $K^f : K^Y \to K^X$ is bijective on connected components whenever K is fibrant.*

The unconditional proof of the fibration extension property for the fibration weak factorization system will be given in Corollary 9.7 of Chap. 9. It uses the equivalence extension property (Chap. 8), which in turn employs a universal fibration (Chap. 7) and the Frobenius condition (Chap. 6). We turn first to the last of these prerequisites.

Chapter 6
The Frobenius Condition

In this chapter, we show that the (unbiased) fibration weak factorization system on the category cSet from Chap. 4 satisfies what has been called the *Frobenius condition*: the left maps are stable under pullback along the right maps (see [72]). This will imply the *right properness* of our model structure: the weak equivalences are preserved by pullback along fibrations. In the present setting, it then follows that the entire model structure is stable under such a base change. The Frobenius condition will be used in the proof of the equivalence extension property in Chap. 8.

A proof of Frobenius in the related setting of cubical sets *with connections* was given in [35] using conventional, functorial methods. By contrast, the type theoretic approach of [28] provides a proof that is much more direct, and can also be modified to work without connections (as in [5]). That approach proves the dual fact that the *pushforward* operation, which is right adjoint to pullback and always exists in a topos, preserves fibrations when applied along a fibration. This corresponds to the type-theoretic Π-formation rule, and the proof given in op.cit. is entirely in type theory. It also employs a reduction of box filling (in all dimensions) to an apparently weaker condition of *Kan composition* (in all dimensions), which merely "puts a lid on" the open box, rather than filling it. This aspect of the type theoretic proof can also be described functorially, but is not used in the proof given here, and will therefore not be discussed further (see [53] for a description of Kan composition with connections, and [8] for the same without connections). See [40] and especially [16] for recent, improved proofs of the Frobenius condition for unbiased fibrations.

© The Author(s), under exclusive license to Springer Nature Switzerland AG 2026
S. Awodey, *Cartesian Cubical Model Categories*, Lecture Notes
in Mathematics 2385, https://doi.org/10.1007/978-3-032-08730-0_6

6.1   From Biased to Unbiased

Our proof takes the approach that was used to determine the unbiased fibrations, namely we first establish the result in the *biased but generic* setting, and then transfer it to the unbiased setting by pulling back along the base change $\mathsf{cSet} \to \mathsf{cSet}/_I$. We first give the second step as a conditional statement.

Proposition 6.1 *Suppose the δ-biased fibrations in $\mathsf{cSet}/_I$ satisfy the Frobenius condition. Then the unbiased fibrations in cSet also satisfy the Frobenius condition.*

Proof This follows almost immediately from the fact that the pullback functor $I^* : \mathsf{cSet} \to \mathsf{cSet}/_I$ preserves the locally cartesian closed structure, takes unbiased fibrations to δ-biased ones, and reflects δ-biased fibrations to unbiased ones. In detail, let unbiased fibrations $B \twoheadrightarrow A$ and $A \twoheadrightarrow X$ in cSet be given, and we wish to find $C \twoheadrightarrow X$ and $e : A \times_X C \to B$ over A, universal in the way recalled in the diagram below.

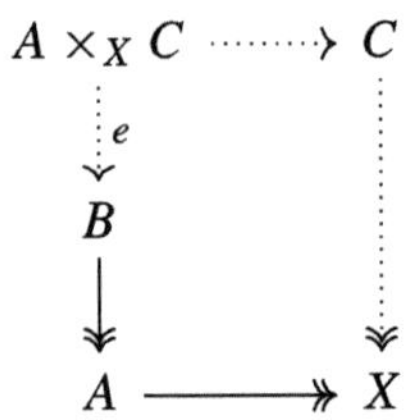

$$
\begin{array}{ccc}
A \times_X C & \dashrightarrow & C \\
\;\downarrow{\scriptstyle e} & & \vdots \\
B & & \vdots \\
\downarrow & & \vdots \\
A & \twoheadrightarrow & X
\end{array}
\tag{6.1.1}
$$

Take the pushforward $C := A_* B \to X$, and its associated map $e : A \times_X C \to B$, in the locally cartesian closed category cSet. Since fibrations are stable under (all) pullbacks, it then suffices to show that $C \to X$ is a fibration.

By definition, $C \to X$ is an unbiased fibration in cSet just in case the base change $I^*C \to I^*X$ is a δ-biased fibration in the slice category $\mathsf{cSet}/_I$. Since the pullback functor $I^* : \mathsf{cSet} \to \mathsf{cSet}/_I$ preserves all lcc structure, over I^*X we have an iso,

$$
I^*C = I^*(A_* B) \cong (I^*A)_* I^* B \,,
$$

where the pushforward $(I^*A)_* I^* B$ is taken in the topos $\mathsf{cSet}/_I$. But $I^*B \to I^*A$ and $I^*A \to I^*X$ are δ-biased fibrations in $\mathsf{cSet}/_I$ because $B \to A$ and $A \to X$ were assumed to be unbiased fibrations in cSet. Since we are assuming the Frobenius condition for δ-biased fibrations in $\mathsf{cSet}/_I$, the pushforward $I^*C \cong (I^*A)_* I^* B \to I^*X$ is also a δ-biased fibration, as required. $\square$

6.2   Frobenius for δ-Biased Fibrations

The results proved in this chapter will be applied to the slice category $\mathsf{cSet}/_I$ and the generic point $\delta : 1 \to I = I^*I$, and then used to infer the Frobenius condition for unbiased fibrations in cSet via Proposition 6.1. But since nothing depends on this particular case, we shall simply assume a pointed object $\delta : 1 \to I$ in an arbitrary topos $\mathcal{E}$, and prove the result for δ-biased fibrations in $\mathcal{E}$. (Indeed, in this chapter $\mathcal{E}$ may even be just a locally cartesian closed category with a class of Cartesian cofibrations in the sense of Appendix A.)

Recall from Definition 4.6 that a map $f : A \to X$ is a δ-biased fibration just if the map $\delta \Rightarrow f$ admits a relative +-algebra structure, and is therefore a trivial fibration. The definition of the pullback-hom $\delta \Rightarrow f$ is recalled below.

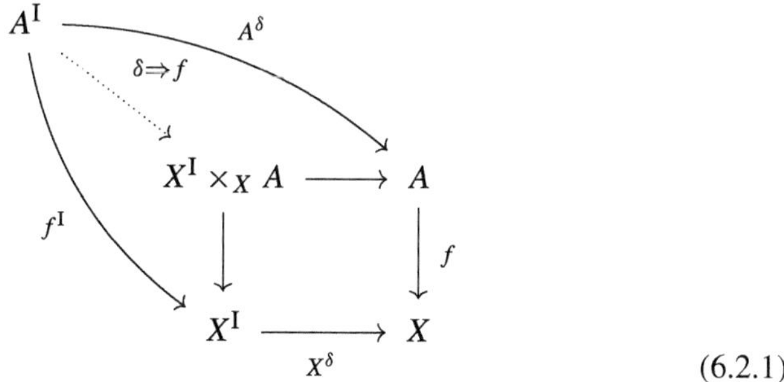

$$\tag{6.2.1}$$

Let us write this condition schematically as follows:

$$\tag{6.2.2}$$

where $\epsilon = X^\delta$ and $A_\epsilon = X^I \times_X A$, and the struck-through arrow indicates that it admits a +-algebra structure.

Lemma 6.2 *Let $A \to X$ be a δ-biased fibration and $t : Y \to X$ any map, then the pullback $t^* A \to Y$ is also a δ-biased fibration.*

Proof This is clear from the fact that the δ-biased fibrations can be made into the right class of a weak factorization system (by reasoning analogous to that for Proposition 4.20), but it will be useful to see how the structure indicated in (6.2.2) is itself stable under pullback. Indeed, consider the following commutative diagram, in which the front face of the cube is the pullback in question, and the right and left sides are the respective versions of the construction in (6.2.2).

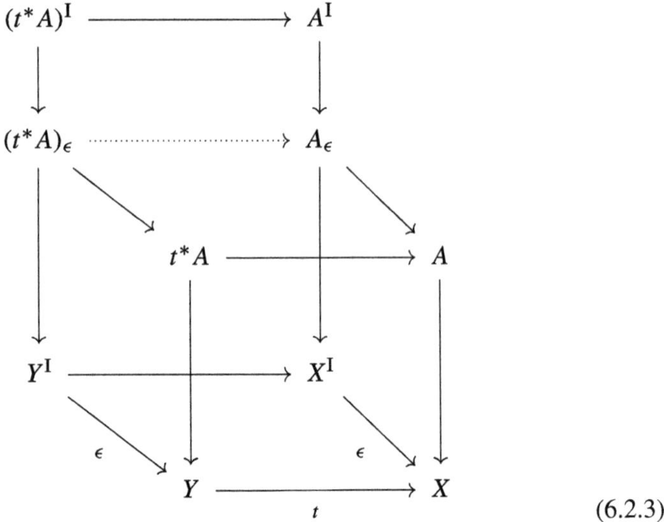

$$(6.2.3)$$

The rear square of solid arrows is the image of the front face under the pathobject functor and is therefore also a pullback. The base commutes by the naturality of the maps ϵ, as does a corresponding top square involving further such ϵ's not shown. Note that these naturality squares need not be pullbacks, but the vertical squares on the sides are, by construction. It follows that there is a dotted arrow as shown, making the resulting lower rear square commute. That lower square is then also a pullback, since the other vertical faces of the resulting cube are pullbacks, and thus finally, the upper rear square is also a pullback.

Now if $A \to X$ is a δ-biased fibration, then $A^{\mathrm{I}} \to A_\epsilon$ is a trivial fibration, and then so is its pullback $(t^*A)^{\mathrm{I}} \to (t^*A)_\epsilon$ since relative $+$-algebras are stable under pullback. Therefore the pullback $t^*A \to Y$ is also a δ-biased fibration. $\qquad\square$

Remark 6.3 In this way we can show algebraically that the pullback of a δ-biased fibration is again one by pulling back the structure that makes it so. In Sect. 7.3, the pullback stability of the fibration structure will be used in the construction of a universal fibration via a closely related argument.

Lemma 6.4 *Let* $\alpha \: : \: A \to X$ *and* $\beta \: : \: B \to A$ *be* δ-*biased fibrations, then the composite* $\alpha \circ \beta : B \to X$ *is also a* δ-*biased fibration.*

Proof Again for maps in the right class of a weak factorization system this is immediate. But let us see how the fibration structures also compose. We have the following diagram for the fibration structures on $B \to A$ and $A \to X$ (with obvious notation).

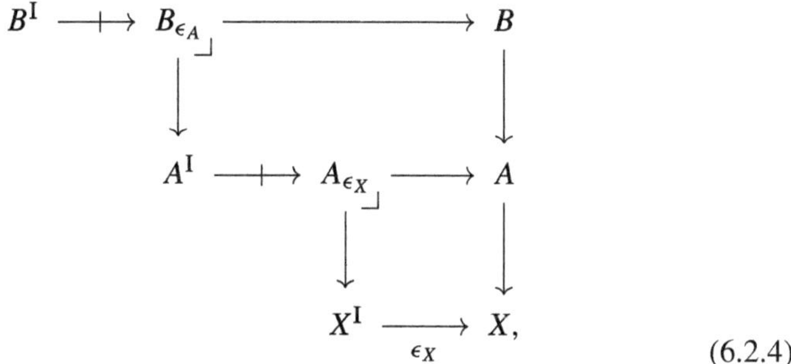

$$\tag{6.2.4}$$

Pulling back $B \to A$ in two steps we therefore obtain the intermediate map $B_{\epsilon_X} \to A_{\epsilon_X}$ indicated in the following diagram.

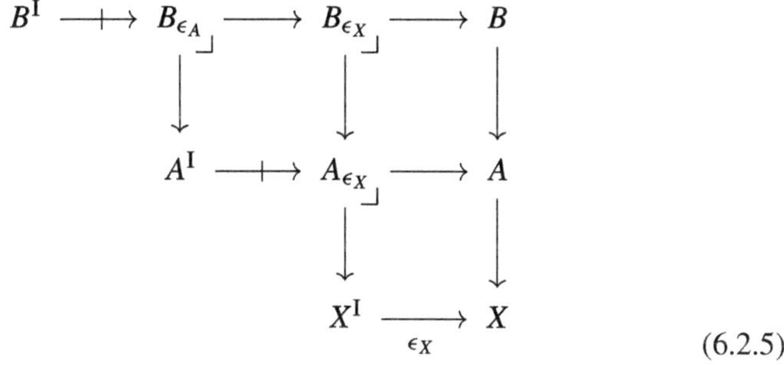

$$\tag{6.2.5}$$

Now use the fact that a trivial fibration structure (i.e. a +-algebra structure) has a canonical pullback along any map, and that two such structures have a canonical composition (cf. Remark 3.11), to obtain a trivial fibration structure for the indicated composite map $B^I \to B_{\epsilon_X}$, which is then a fibration structure for the composite $B \to A \to X$. $\square$

Proposition 6.5 (δ-Biased Frobenius)   *If $\alpha : A \to X$ and $\beta : B \to A$ are δ-biased fibrations, then the pushforward $\alpha_* \beta : \Pi_A B \to X$ is also a δ-biased fibration.*

Proof   Given δ-biased fibrations $\alpha : A \to X$ and $\beta : B \to A$, let $a : A^I \to A_\epsilon$ and $b : B^I \to a^* B_\epsilon$ be the associated trivial fibrations, so that we have the situation of diagram (6.2.5), with all three squares pullbacks.

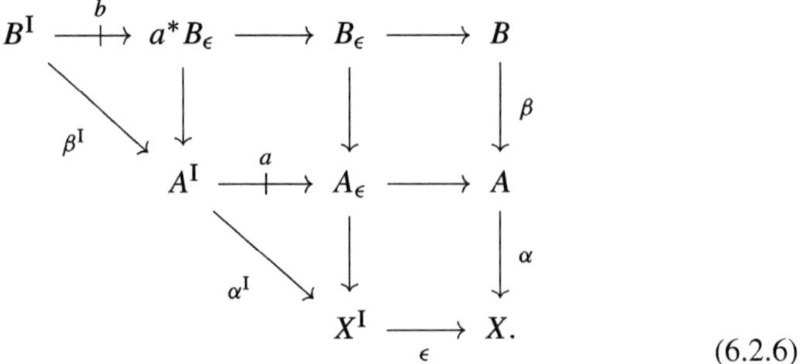

$$(6.2.6)$$

Taking the pushforward of the righthand vertical column gives a map,

$$\gamma := \alpha_* \beta : \Pi_A B \to X\,,$$

and placing it underneath, along with the corresponding construction from (6.2.2), we then have the following commutative diagram.

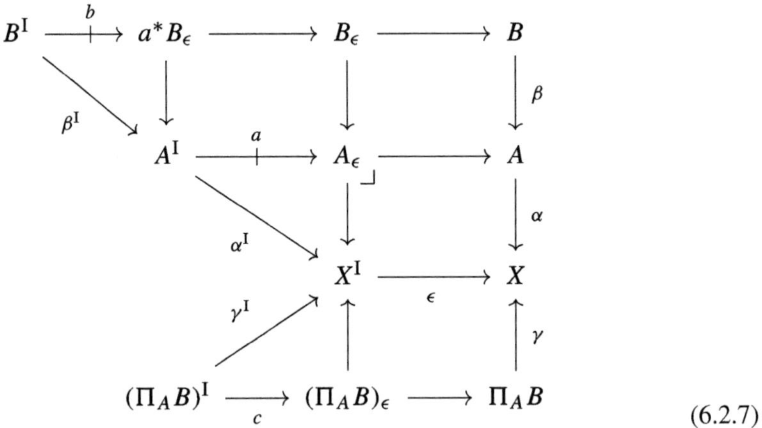

$$(6.2.7)$$

We wish to show that the indicated map $c : (\Pi_A B)^{\mathrm{I}} \to (\Pi_A B)_\epsilon$ admits a +-algebra structure. This we will do by showing that it is a retract of a known +-algebra. Namely, we can apply the pushforward along the map $\alpha^{\mathrm{I}} : A^{\mathrm{I}} \to X^{\mathrm{I}}$ to the +-algebra $b : B^{\mathrm{I}} \to a^* B_\epsilon$ regarded as an arrow over A^{I}. We obtain an arrow over X^{I} of the form

$$\Pi_{A^{\mathrm{I}}} b : \Pi_{A^{\mathrm{I}}} B^{\mathrm{I}} \longrightarrow \Pi_{A^{\mathrm{I}}} a^* B_\epsilon \tag{6.2.8}$$

which is indeed a +-algebra, since these are preserved under pushing forward, by Remark 3.11.

Next, observe that by the Beck-Chevalley condition for the indicated central pullback in (6.2.7), for the codomain of $c : (\Pi_A B)^{\mathrm{I}} \to (\Pi_A B)_\epsilon$ we have an isomorphism

$$(\Pi_A B)_\epsilon \;\cong\; \Pi_{A_\epsilon} B_\epsilon \qquad \text{over } X^{\mathrm{I}}.$$

And since $\Pi_{A^{\mathrm{I}}} \cong \Pi_{A_\epsilon} \circ a_*$, for the codomain of our $+$-algebra $\Pi_{A^{\mathrm{I}}} b$ from (6.2.8) we also have

$$\Pi_{A^{\mathrm{I}}} a^* B_\epsilon \;\cong\; \Pi_{A_\epsilon} a_* a^* B_\epsilon \,.$$

Thus the image of the unit $\eta : B_\epsilon \to a_* a^* B_\epsilon$ under Π_{A_ϵ} provides a map $\sigma := \Pi_{A_\epsilon}\eta$ over X^{I} of the form:

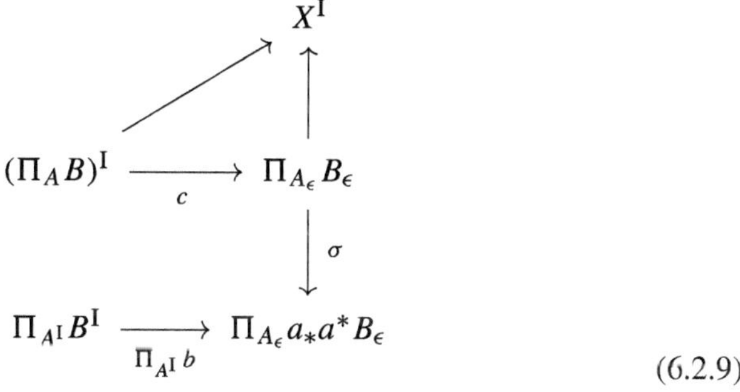

$$\tag{6.2.9}$$

Our goal is now to determine further arrows φ, ψ, τ as indicated below, exhibiting c as a retract of $\Pi_{A^{\mathrm{I}}} b$ in the arrow category over X^{I}.

$$\tag{6.2.10}$$

- For φ, we require a map

$$\varphi : (\Pi_A B)^I \to \Pi_{A^I} B^I \qquad \text{over } X^I.$$

Consider the following diagram, which is based on (6.2.2).

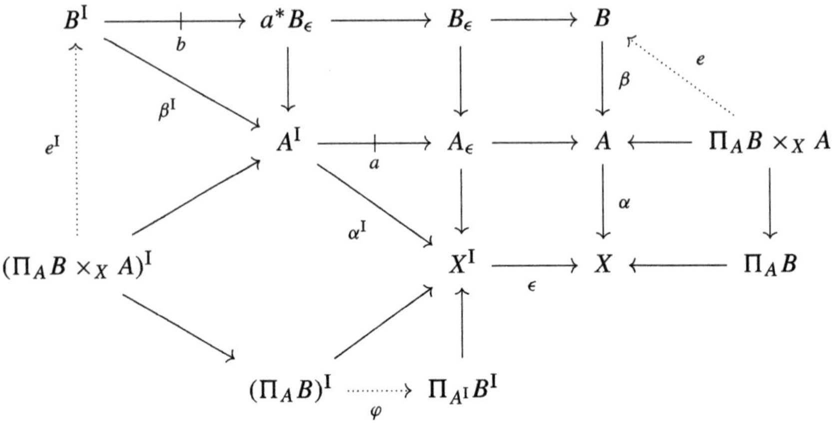

$$(6.2.11)$$

The map e is the counit at $\beta : B \to A$ of the pullback-pushforward adjunction along $\alpha : A \to X$. The right-hand side of the diagram, including e and the associated pullback square, reappears (mirrored) on the left under the functor $(-)^I$, which preserves the pullback. Thus we can take φ to be the transpose of e^I under the pullback-pushforward adjunction along $\alpha^I : A^I \to X^I$,

$$\varphi := \widetilde{e^I}.$$

An easy diagram chase involving the pullback-pushforward adjunction along $A_\epsilon \to X^I$ shows that the upper square in (6.2.10) then commutes.

- For τ: referring to the diagram (6.2.11), since $a : A^I \to A_\epsilon$ is a trivial fibration, it has a section $o : A_\epsilon \to A^I$ by Lemma 3.10. Pulling $a^* B_\epsilon \to A^I$ back along o results in an iso,

$$o^* a^* B_\epsilon \cong B_\epsilon \quad \text{over } A_\epsilon$$

and so by the adjunction $o^* \dashv o_*$ there is an associated map,

$$a^* B_\epsilon \to o_* B_\epsilon \quad \text{over } A^I$$

to which we can apply a_* to obtain a map,

$$t : a_* a^* B_\epsilon \to a_* o_* B_\epsilon \cong B_\epsilon \quad \text{over } A_\epsilon .$$

This map t is evidently a retraction of the unit $\eta : B_\epsilon \to a_* a^* B_\epsilon$ over A_ϵ. Applying the functor Π_{A_ϵ} therefore gives the desired retraction of σ,

$$\tau := \Pi_{A_\epsilon} t : \Pi_{A_\epsilon} a_* a^* B_\epsilon \to \Pi_{A_\epsilon} B_\epsilon .$$

- For ψ, we require a map

$$\psi : \Pi_{A^I} B^I \to (\Pi_A B)^I \qquad \text{over } X^I.$$

Consider the following diagram resulting from combining (6.2.2) and (6.2.10), in which all solid arrows are those already introduced. The dotted arrow labelled p is the evident composite.

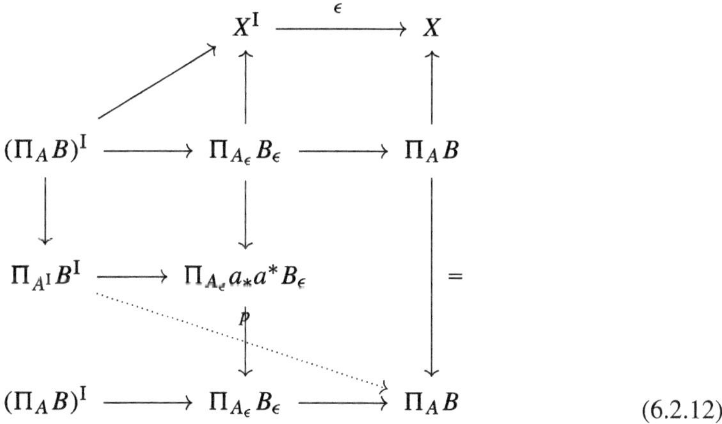

$$(6.2.12)$$

The lower horizontal composite is the evaluation of the pathobject $(\Pi_A B)^I$ at the point $\delta : 1 \to I$,

$$\epsilon_{\Pi_A B} = (\Pi_A B)^\delta : (\Pi_A B)^I \longrightarrow (\Pi_A B)^1 \cong \Pi_A B .$$

This is constructed from the (cartesian closed) evaluation,

$$\mathsf{eval} : I \times (\Pi_A B)^I \longrightarrow \Pi_A B$$

which is the counit of $I \times (\,\_\,) \dashv (\,\_\,)^I$, as the composite shown below.

$$
\begin{array}{ccc}
(\Pi_A B)^I & \xrightarrow{\;\;\epsilon_{\Pi_A B}\;\;} & \Pi_A B \\[2mm]
\cong \Big\downarrow & & \Big\uparrow \mathsf{eval} \\[2mm]
1 \times (\Pi_A B)^I & \xrightarrow[\;\delta \times (\Pi_A B)^I\;]{} & I \times (\Pi_A B)^I
\end{array}
$$

$$(6.2.13)$$

Let us analyse this evaluation at δ further, in terms of the *locally* cartesian closed structure associated to the base changes along the section $\delta : 1 \to I$ and retraction $I \to 1$ in $\mathcal{E}$. Since $\mathrm{id} \cong \delta^*I^* : \mathcal{E} \to \mathcal{E}/_I \to \mathcal{E}$, the map $\epsilon_{\Pi_A B}$ can be rewritten as follows.

$$
\begin{array}{ccc}
(\Pi_A B)^I & \xrightarrow{\ \epsilon_{\Pi_A B}\ } & \Pi_A B \\[2mm]
\cong \big\downarrow & & \big\downarrow \cong \\[2mm]
\delta^*I^*((\Pi_A B)^I) & \xrightarrow{\ \delta^*I^*\epsilon_{\Pi_A B}\ } & \delta^*I^*\Pi_A B \\[2mm]
\cong \big\downarrow & & \big\downarrow = \\[2mm]
\delta^*I^*I_*I^*\Pi_A B & \xrightarrow[\ \delta^*\varepsilon\ ]{} & \delta^*I^*\Pi_A B
\end{array}
\qquad (6.2.14)
$$

where the map $\delta^*\varepsilon$ across the bottom is the counit of the adjunction $I^* \dashv I_*$, taken at $I^*\Pi_A B$, and then pulled back along $\delta : 1 \to I$. Before taking the pullback, we therefore have the following iso over I between that counit ε_{I^*} and the image under I^* of the previously considered evaluation $\epsilon : (\Pi_A B)^I \to \Pi_A B$ from (6.2.13).

$$
\begin{array}{ccc}
I^*((\Pi_A B)^I) & \xrightarrow{\ I^*\epsilon\ } & I^*\Pi_A B \\[2mm]
\cong \big\downarrow & & \big\downarrow = \\[2mm]
I^*I_*I^*\Pi_A B & \xrightarrow[\ \varepsilon_{I^*}\ ]{} & I^*\Pi_A B .
\end{array}
\qquad (6.2.15)
$$

Now let us apply I^* to (6.2.12) to get the map I^*p in the diagram below, which therefore factors (up to (6.2.15)) through the counit ε_{I^*} as $\varepsilon_{I^*} \circ I^*(\widetilde{I^*p})$, where $\widetilde{I^*p}$ is the adjoint transpose of I^*p, as shown.

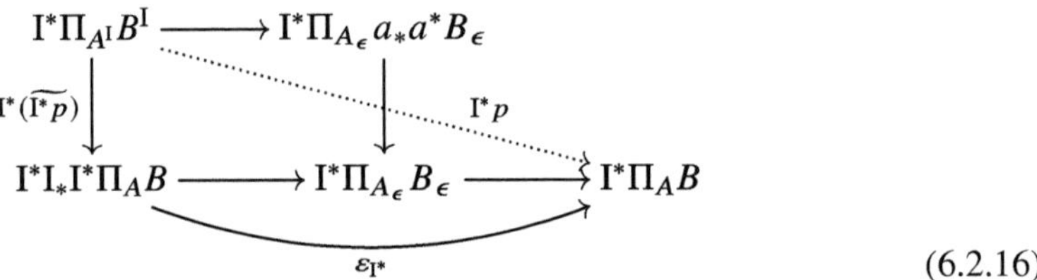

$$
\qquad (6.2.16)
$$

We can therefore set

$$
\psi := \widetilde{I^*p}\,,
$$

and we obtain $\epsilon \circ \psi = p$, from which it follows that the square in (6.2.16) commutes by the definition of $\Pi_{A_\epsilon} B_\epsilon$ as a pullback. The same square without I^* then also commutes by applying the retraction δ^*.

We have now defined all the maps indicated below, the squares involving φ and ψ commute, and the composite of σ and τ is the identity.

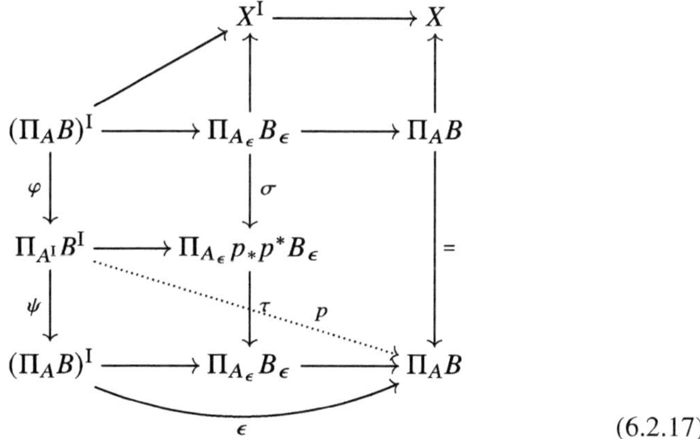

$$\tag{6.2.17}$$

To see that $\psi \circ \varphi = 1$, an easy chase through the diagram (6.2.17) shows that

$$\epsilon \cup \psi \cup \varphi = p \circ \varphi = \epsilon .$$

Thus by applying I* and using (6.2.15) we have $\varepsilon_{I*} \circ \mathrm{I}^*(\psi \circ \varphi) = \varepsilon_{I*}$, and so $\psi \circ \varphi = \bar{\varepsilon_{I*}} = 1$. $\square$

From Proposition 6.1 we then have:

Corollary 6.6 (Unbiased Frobenius)   *The unbiased fibration weak factorization system on* cSet *satisfies the Frobenius condition.*

Corollary 6.7   *Unbiased fibrations are closed under pushforward along unbiased fibrations. Thus given unbiased fibrations $X \twoheadrightarrow Z$ and $Y \twoheadrightarrow Z$ over any base Z, the relative exponential $Y^X = X_* X^* Y \twoheadrightarrow Z$, formed in the slice over Z, is again an unbiased fibration.*

Remark 6.8   We note in passing that the proof just given for the δ-biased case of Frobenius, Proposition 6.5, made no use of the fact that $\delta : 1 \to \mathrm{I}$ is generic, nor even that we were working in the slice category over I. Indeed the same algebraic argument works for p-biased fibrations for any point $p : 1 \to \mathrm{I}$ of any object I, in any (quasi-)topos $\mathcal{E}$.

Chapter 7
A Universal Fibration

We shall construct a *universal small fibration* $\dot{\mathcal{U}} \to \mathcal{U}$, which is a classifier for small fibrations in cSet. It will be shown in Chap. 9 that the base object $\mathcal{U}$ is fibrant, using the fact to be proved in Chap. 8 that the map $\dot{\mathcal{U}} \to \mathcal{U}$ itself is *univalent*, in a sense to be made precise.

Our construction of $\dot{\mathcal{U}} \to \mathcal{U}$ makes use, first of all, of a new description of the well-known Hofmann-Streicher universe in any category $\widehat{\mathbb{C}} = [\mathbb{C}^{\mathrm{op}}, \mathsf{Set}]$ of presheaves on a small category $\mathbb{C}$, which was used in [42] to interpret dependent type theory. Some of the material in this chapter was published in preliminary form as [10].

7.1 Classifying Families

Definition 7.1 ([42]) Let $\mathbb{C}$ be a small category. A (type-theoretic) *universe* (U, El) consists of $U \in \widehat{\mathbb{C}}$ and $\mathsf{El} \in \widehat{\int_{\mathbb{C}} U}$ with:

$$U(c) = \mathsf{Cat}(\mathbb{C}/_c{}^{\mathrm{op}}, \mathsf{Set}) \tag{7.1.1}$$

$$\mathsf{El}(c, A) = A(id_c) \tag{7.1.2}$$

with the evident associated action on morphisms.

A few comments are required:

- In contrast to op.cit., in (7.1.1) we take the underlying set of objects of the functor category $\widehat{\mathbb{C}/_c} = [\mathbb{C}/_c{}^{\mathrm{op}}, \mathsf{Set}]$.
- As in ibid., (7.1.2) adopts the "categories with families" point of view in describing an arrow $E \to U$ in $\widehat{\mathbb{C}}$ equivalently as a presheaf on the category of

© The Author(s), under exclusive license to Springer Nature Switzerland AG 2026
S. Awodey, *Cartesian Cubical Model Categories*, Lecture Notes
in Mathematics 2385, https://doi.org/10.1007/978-3-032-08730-0_7

elements $\int_{\mathbb{C}} U$, using

$$\widehat{\mathbb{C}}/U \simeq \widehat{\int_{\mathbb{C}} U} \tag{7.1.3}$$

where

$$E(c) = \coprod_{A \in U(c)} \mathsf{E}l(c, A).$$

The argument $(c, A) \in \int_{\mathbb{C}} U$ in (7.1.2) thus consists of an object $c \in \mathbb{C}$ and an element $A \in U(c)$.

- To account for size issues, the authors of ibid. assume a Grothendieck universe u in Set, the elements of which are called *small*. The category $\mathbb{C}$ is assumed to be small, as are the values of the presheaves, unless otherwise stated.

The presheaf U, which is not small, is then regarded as the Grothendieck universe u "lifted" from Set to $[\mathbb{C}^{\mathrm{op}}, \mathsf{Set}]$. We first analyse this specification of $(U, \mathsf{E}l)$ from a different perspective, in order to establish its basic property as a classifier for small families in $\widehat{\mathbb{C}}$.

7.1.1   A Realization-Nerve Adjunction

For a presheaf X on $\mathbb{C}$, recall that the category of elements is the comma category,

$$\int_{\mathbb{C}} X = \mathsf{y}_{\mathbb{C}}/X ,$$

where $\mathsf{y}_{\mathbb{C}} : \mathbb{C} \to \mathsf{Set}^{\mathbb{C}^{\mathrm{op}}}$ is the Yoneda embedding, which we sometimes suppress and write simply $\mathbb{C}/X$ for $\mathsf{y}_{\mathbb{C}}/X$.

Proposition 7.2 ([39], §28) *The category of elements functor*

$$\int_{\mathbb{C}} : \widehat{\mathbb{C}} \longrightarrow \mathsf{Cat}$$

has a right adjoint,

$$v_{\mathbb{C}} : \mathsf{Cat} \longrightarrow \widehat{\mathbb{C}} .$$

For a small category $\mathbb{A}$, we shall call the presheaf $v_{\mathbb{C}}(\mathbb{A})$ the $(\mathbb{C}\text{-})$nerve of $\mathbb{A}$.

Proof The adjunction $\int_{\mathbb{C}} \dashv v_{\mathbb{C}}$ is an instance of the usual "realization/nerve" adjunction, here with respect to the covariant slice category functor $\mathbb{C}/- : \mathbb{C} \to \mathsf{Cat}$, as indicated below.

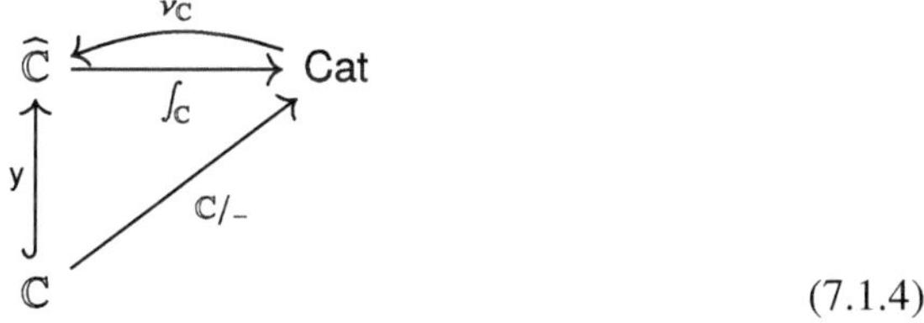

$$(7.1.4)$$

In detail, for $\mathbb{A} \in \mathsf{Cat}$ and $c \in \mathbb{C}$, let $\nu_{\mathbb{C}}(\mathbb{A})(c)$ be the Hom-set of functors,

$$\nu_{\mathbb{C}}(\mathbb{A})(c) = \mathsf{Cat}\big(\mathbb{C}/c \, , \, \mathbb{A}\big) \, ,$$

with contravariant action on $h \, : \, d \to c$ given by pre-composing a functor $P \, : \, \mathbb{C}/c \to \mathbb{A}$ with the post-composition functor

$$\mathbb{C}/h : \mathbb{C}/d \longrightarrow \mathbb{C}/c \, .$$

For the adjunction, observe that the slice category $\mathbb{C}/c$ is the category of elements of the representable functor $\mathsf{y}c$,

$$\textstyle\int_{\mathbb{C}} \mathsf{y}c \; \cong \; \mathbb{C}/c \, .$$

Thus for representables $\mathsf{y}c$, we have the required natural isomorphism

$$\widehat{\mathbb{C}}\big(\mathsf{y}c \, , \, \nu_{\mathbb{C}}(\mathbb{A})\big) \; \cong \; \nu_{\mathbb{C}}(\mathbb{A})(c) \; = \; \mathsf{Cat}\big(\mathbb{C}/c \, , \, \mathbb{A}\big) \; \cong \; \mathsf{Cat}\big(\textstyle\int_{\mathbb{C}} \mathsf{y}c \, , \, \mathbb{A}\big) \, .$$

For arbitrary presheaves X, one uses the presentation of X as a colimit of representables over the index category $\int_{\mathbb{C}} X$, and the easy to prove fact that $\int_{\mathbb{C}}$ itself preserves colimits. Indeed, for any category $\mathbb{D}$, we have an isomorphism in Cat,

$$\varinjlim_{d \in \mathbb{D}} \mathbb{D}/d \; \cong \; \mathbb{D} \, .$$

$\square$

When $\mathbb{C}$ is fixed, we may omit the subscript in the notation $\mathsf{y}_{\mathbb{C}}$ and $\int_{\mathbb{C}}$ and $\nu_{\mathbb{C}}$. The unit and counit maps of the adjunction $\int \dashv \nu$,

$$\eta : X \longrightarrow \nu \textstyle\int X \, ,$$

$$\epsilon : \textstyle\int \nu \mathbb{A} \longrightarrow \mathbb{A} \, ,$$

are then as follows. At $c \in \mathbb{C}$, for $x : \mathsf{y}c \to X$, the functor $(\eta_X)_c(x) : \mathbb{C}/c \to \mathbb{C}/X$ is just composition with x,

$$(\eta_X)_c(x) = \mathbb{C}/x : \mathbb{C}/c \longrightarrow \mathbb{C}/X \, . \qquad\qquad (7.1.5)$$

For $\mathbb{A} \in$ **Cat**, the functor $\epsilon : \int \nu\mathbb{A} \to \mathbb{A}$ takes a pair $(c \in \mathbb{C}, f : \mathbb{C}/_c \to \mathbb{A})$ to the object $f(1_c) \in \mathbb{A}$,

$$\epsilon(c, f) = f(1_c).$$

Lemma 7.3 *For any $f : Y \to X$, the naturality square below is a pullback.*

$$
\begin{array}{ccc}
Y & \xrightarrow{\eta_Y} & \nu\int Y \\
{\scriptstyle f}\downarrow & & \downarrow{\scriptstyle \nu\int f} \\
X & \xrightarrow{\eta_X} & \nu\int X.
\end{array}
\tag{7.1.6}
$$

Proof It suffices to prove this for the case $f : X \to 1$. Thus consider the square

$$
\begin{array}{ccc}
X & \xrightarrow{\eta_X} & \nu\int X \\
\downarrow & & \downarrow \\
1 & \xrightarrow{\eta_1} & \nu\int 1.
\end{array}
\tag{7.1.7}
$$

Evaluating at $c \in \mathbb{C}$ and applying (7.1.5) gives the following square in **Set**.

$$
\begin{array}{ccc}
Xc & \xrightarrow{\mathbb{C}/_-} & \mathsf{Cat}(\mathbb{C}/_c\,,\,\mathbb{C}/_X) \\
\downarrow & & \downarrow \\
1c & \xrightarrow[\mathbb{C}/_-]{} & \mathsf{Cat}(\mathbb{C}/_c\,,\,\mathbb{C}/_1)
\end{array}
\tag{7.1.8}
$$

The image of $* \in 1c$ along the bottom is the forgetful functor $U_c : \mathbb{C}/_c \to \mathbb{C}$, and its fiber under the map on the right is the set of functors $F : \mathbb{C}/_c \to \mathbb{C}/_X$ such that $U_X \circ F = U_c$, where $U_X : \mathbb{C}/_X \to \mathbb{C}$ is also a forgetful functor. But any such F is uniquely of the form $\mathbb{C}/_x$ for $x = F(1_c) : yc \to X$. $\qquad\square$

7.1.2 A Universal Family

For the terminal presheaf $1 \in \widehat{\mathbb{C}}$ we have an iso $\int 1 \cong \mathbb{C}$, so for every $X \in \widehat{\mathbb{C}}$ there is a canonical projection $\int X \to \mathbb{C}$, which is a discrete fibration. It follows that for any map $Y \to X$ of presheaves, the associated map $\int Y \to \int X$ is also a discrete fibration. Ignoring size issues temporarily, recall that discrete fibrations in **Cat** are classified by the forgetful functor $\mathsf{Set}_*^{\mathrm{op}} \to \mathsf{Set}^{\mathrm{op}}$ from (the opposites of) the category of pointed sets to that of sets (cf. [74]). For every presheaf $X \in \widehat{\mathbb{C}}$, we therefore have a pullback diagram in Cat,

$$
\begin{array}{ccc}
\int X & \longrightarrow & \dot{\mathsf{Set}}^{\mathrm{op}} \\
\downarrow & \lrcorner & \downarrow \\
\mathbb{C} & \xrightarrow{\quad X \quad} & \mathsf{Set}^{\mathrm{op}}.
\end{array}
\tag{7.1.9}
$$

Using $\mathbb{C} \cong \int 1$ and transposing by the adjunction $\int \dashv \nu$ then gives a commutative square in $\widehat{\mathbb{C}}$ of the form:

$$
\begin{array}{ccc}
X & \longrightarrow & \nu\dot{\mathsf{Set}}^{\mathrm{op}} \\
\downarrow & & \downarrow \\
1 & \xrightarrow{\quad \tilde{X} \quad} & \nu\mathsf{Set}^{\mathrm{op}}.
\end{array}
\tag{7.1.10}
$$

Lemma 7.4 *The square* (7.1.10) *is a pullback in* $\widehat{\mathbb{C}}$. *More generally, for any map* $Y \to X$ *in* $\widehat{\mathbb{C}}$, *there is a canonical pullback square*

$$
\begin{array}{ccc}
Y & \longrightarrow & \nu\dot{\mathsf{Set}}^{\mathrm{op}} \\
\downarrow & \lrcorner & \downarrow \\
X & \longrightarrow & \nu\mathsf{Set}^{\mathrm{op}}.
\end{array}
\tag{7.1.11}
$$

Proof Apply the right adjoint ν to the pullback square (7.1.9) and paste the naturality square (7.1.6) from Lemma 7.3 on the left, to obtain the transposed square (7.1.11) as a pasting of two pullbacks. $\qquad\qquad\square$

Let us write $\dot{\mathcal{V}} \to \mathcal{V}$ for the vertical map on the right in (7.1.11), setting

$$
\dot{\mathcal{V}} := \nu\dot{\mathsf{Set}}^{\mathrm{op}}
\tag{7.1.12}
$$

$$
\mathcal{V} := \nu\mathsf{Set}^{\mathrm{op}}.
$$

We summarize our results so far as follows.

Proposition 7.5 *The nerve* $\dot{\mathcal{V}} \to \mathcal{V}$ *of the classifier for discrete fibrations* $\dot{\mathsf{Set}}^{\mathrm{op}} \to \mathsf{Set}^{\mathrm{op}}$, *as defined in* (7.1.12), *classifies natural transformations* $Y \to X$ *in* $\widehat{\mathbb{C}}$, *in the sense that there is always a pullback square,*

$$
\begin{array}{ccc}
Y & \longrightarrow & \dot{\mathcal{V}} \\
\downarrow & \lrcorner & \downarrow \\
X & \xrightarrow{\quad Y \quad} & \mathcal{V}.
\end{array}
\tag{7.1.13}
$$

The classifying map $\tilde{Y} : X \to \mathcal{V}$ is determined by the adjunction $\int \dashv \nu$ as the transpose of the classifying map of the discrete fibration $\int Y \to \int X$.

Given a natural transformation $Y \to X$, the classifying map $\tilde{Y} : X \to \mathcal{V}$ is of course not in general unique. Nonetheless, we can use the construction of $\dot{\mathcal{V}} \to \mathcal{V}$ as the nerve of the discrete fibration classifier $\dot{\mathsf{Set}}^{\mathrm{op}} \to \mathsf{Set}^{\mathrm{op}}$, for which classifying functors $\mathbb{C} \to \mathsf{Set}^{\mathrm{op}}$ are unique up to natural isomorphism, to infer the following proposition, which will be required below (cf. [38, 67]). Specifically it is used in the Realignment Lemma for fibrations 7.22, which is needed to prove the fibrancy of the universe, Proposition 9.4.

Proposition 7.6 (Realignment for Families) *Given a monomorphism $c : C \rightarrowtail X$ and a family $Y \to X$, let $y_c : C \to \mathcal{V}$ classify the pullback $c^*Y \to C$. Then there is a classifying map $y : X \to \mathcal{V}$ for $Y \to X$ with $y \circ c = y_c$.*

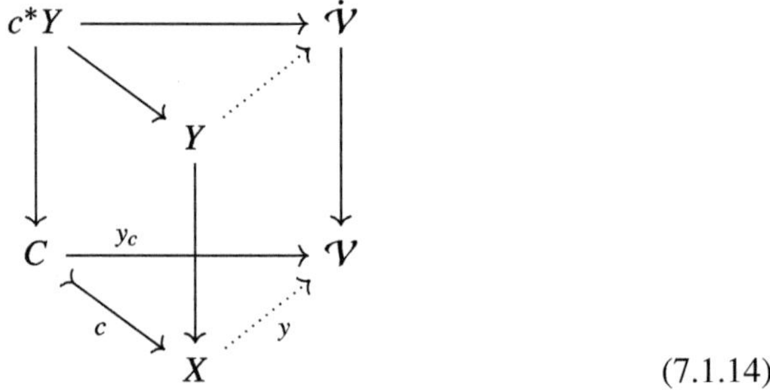

$$(7.1.14)$$

Proof Transposing the realignment problem (7.1.14) for presheaves across the adjunction $\int \dashv \nu$ results in the following realignment problem for discrete fibrations.

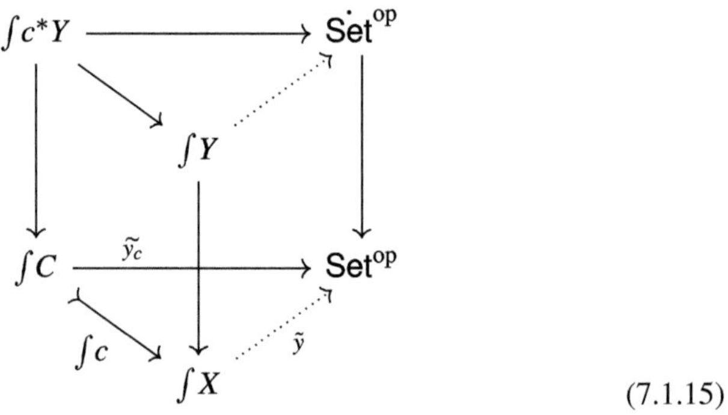

$$(7.1.15)$$

The category of elements functor $\int$ is easily seen to preserve pullbacks, hence monos; thus let us consider the general case of a functor $C : \mathbb{C} \rightarrowtail \mathbb{D}$ which is monic in Cat, a pullback of discrete fibrations as on the left below, and a presheaf $E : \mathbb{C} \to \mathsf{Set}^{\mathrm{op}}$ with $\int E \cong \mathbb{E}$ over $\mathbb{C}$.

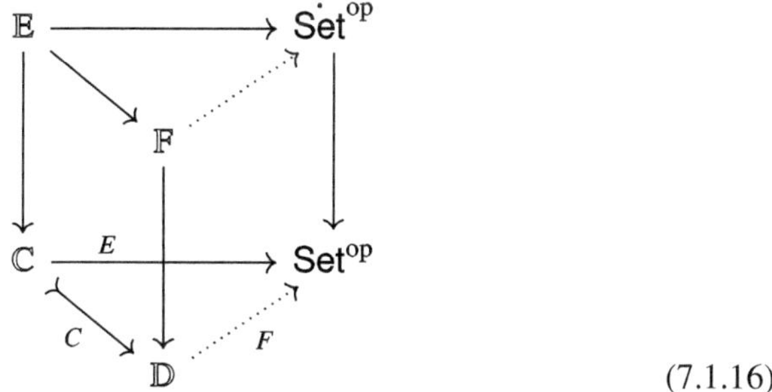

$$(7.1.16)$$

We seek $F : \mathbb{D} \to \mathsf{Set}^{\mathrm{op}}$ with $\int F \cong \mathbb{F}$ over $\mathbb{D}$ and $F \circ C = E$. Let $F_0 : \mathbb{D} \to \mathsf{Set}^{\mathrm{op}}$ with $\int F_0 \cong \mathbb{F}$ over $\mathbb{D}$, which exists since $\mathbb{F} \to \mathbb{D}$ is a discrete fibration. Since $F_0 \circ C$ and E both classify $\mathbb{E}$, there is a natural iso $e : F_0 \circ C \cong E$. Consider the following diagram

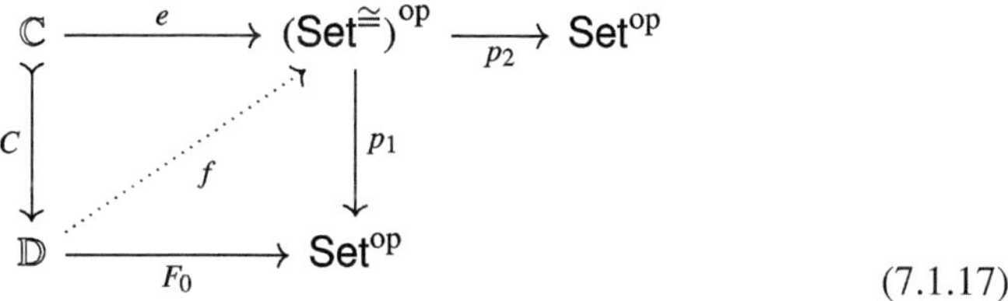

$$(7.1.17)$$

where $\mathsf{Set}^{\cong}$ is the category of isos in Set, with p_1, p_2 the (opposites of the) domain and codomain projections. There is a well-known weak factorization system on Cat (part of the "canonical model structure") with injective-on-objects functors on the left and isofibration equivalences on the right. Thus there is a diagonal filler f as indicated. The functor $F := p_2 \circ f : \mathbb{D} \to \mathsf{Set}^{\mathrm{op}}$ is then the one we seek. $\square$

7.1.3 Small Maps

Of course, as defined in (7.1.12), the classifier $\dot{\mathcal{V}} \to \mathcal{V}$ cannot be a map in $\widehat{\mathbb{C}}$, for reasons of size; we now address this.

For any cardinal number α, call a set α-*small* if its cardinality is strictly less than α. Let $\mathsf{Set}_\alpha \hookrightarrow \mathsf{Set}$ be the full subcategory of α-small sets. Call a map $f : Y \to X$ of presheaves α-*small* if all of the fibers $f_c^{-1}\{x\} \subseteq Yc$ are α-small sets (for all $c \in \mathbb{C}$ and $x \in Xc$). The latter condition is equivalent to saying that for any element $x : yc \to X$, the set of lifts $y : yc \to Y$ of x across f is α-small.

$$(7.1.18)$$

Finally, call a presheaf $X : \mathbb{C}^{\mathrm{op}} \to \mathsf{Set}$ α-*small* if the map $X \to 1$ is α-small. This implies that all of the values Xc are α-small sets, and so the functor $X : \square^{\mathrm{op}} \to \mathsf{Set}$ factors through $\mathsf{Set}_\alpha \hookrightarrow \mathsf{Set}$.

Now let us restrict the specification (7.1.12) of $\dot{\mathcal{V}} \to \mathcal{V}$ to the α-small sets:

$$\dot{\mathcal{V}}_\alpha := \nu \dot{\mathsf{Set}}_\alpha^{\mathrm{op}} \tag{7.1.19}$$

$$\mathcal{V}_\alpha := \nu \mathsf{Set}_\alpha^{\mathrm{op}}.$$

Then the evident forgetful map $\dot{\mathcal{V}}_\alpha \to \mathcal{V}_\alpha$ is a map in the category $\widehat{\mathbb{C}}$ of presheaves, and it is easily seen to be α-small. Moreover, it has the following basic property, which is just a restriction of the basic property of $\dot{\mathcal{V}} \to \mathcal{V}$ stated in Proposition 7.5.

Proposition 7.7 *The map $\dot{\mathcal{V}}_\alpha \to \mathcal{V}_\alpha$ classifies α-small maps $f : Y \to X$ in $\widehat{\mathbb{C}}$, in the sense that there is always a pullback square,*

$$\begin{array}{ccc} Y & \longrightarrow & \dot{\mathcal{V}}_\alpha \\ \downarrow & \lrcorner & \downarrow \\ X & \xrightarrow{\tilde{Y}} & \mathcal{V}_\alpha. \end{array} \tag{7.1.20}$$

The classifying map $\tilde{Y} : X \to \mathcal{V}_\alpha$ is determined by the adjunction $\int \dashv \nu$ as (the factorization of) the transpose of the classifying map of the discrete fibration $\int X \to \int Y$.

Proof If $Y \to X$ is α-small, its classifying map $\tilde{Y} : X \to \mathcal{V}$ factors through $\mathcal{V}_\alpha \hookrightarrow \mathcal{V}$, as indicated below,

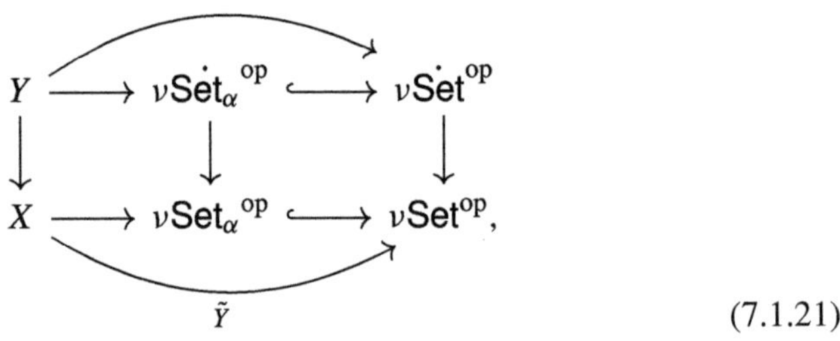

$$\tag{7.1.21}$$

in virtue of the following adjoint transposition,

$$\tag{7.1.22}$$

Note that the square on the right is evidently a pullback, and so the one on the left is, too, because the outer rectangle is the classifying pullback of the discrete fibration $\int Y \to \int X$, as stated. Thus the left square in (7.1.21) is also a pullback. $\qquad\square$

7.1.4 *Examples of Universal Families* $\dot{\mathcal{V}}_\alpha \longrightarrow \mathcal{V}_\alpha$

1. Let $\alpha = \kappa$ a strongly inaccessible cardinal, so that $\mathrm{ob}(\mathsf{Set}_\kappa)$ is a Grothendieck universe. Then the Hofmann-Streicher universe of Definition 7.1 is recovered as the κ-small map classifier

$$E \cong \dot{\mathcal{V}}_\kappa \longrightarrow \mathcal{V}_\kappa \cong U$$

in the sense of Proposition 7.7. Indeed, for $c \in \mathbb{C}$, we have

$$\mathcal{V}_\kappa c \;=\; v(\mathsf{Set}_\kappa^{\mathrm{op}})(c) = \mathsf{Cat}(\mathbb{C}/c,\, \mathsf{Set}_\kappa^{\mathrm{op}}) \;=\; \mathrm{ob}(\widehat{\mathbb{C}/c}) \;=\; Uc. \tag{7.1.23}$$

For $\dot{\mathcal{V}}_\kappa$ we then have,

$$\dot{\mathcal{V}}_\kappa c \;=\; v(\dot{\mathsf{Set}}_\kappa^{\mathrm{op}})(c) \;=\; \mathsf{Cat}(\mathbb{C}/c,\, \dot{\mathsf{Set}}_\kappa^{\mathrm{op}})$$

$$\cong \coprod_{A \in {}^{\scriptscriptstyle'}\mathcal{V}_\kappa c} \mathsf{Cat}_{\mathbb{C}/c}(\mathbb{C}/c,\, A^*\mathsf{Set}_\kappa^{\mathrm{op}}) \tag{7.1.24}$$

where the A-summand in (7.1.24) is defined by taking sections of the pullback indicated below.

$$
\begin{array}{ccc}
A^*\mathsf{Set}_\kappa^{\mathrm{op}} & \longrightarrow & \dot{\mathsf{Set}}_\kappa^{\mathrm{op}} \\[2pt]
\Big\downarrow & \lrcorner \nearrow & \Big\downarrow \\[2pt]
\mathbb{C}/c & \xrightarrow[\;A\;]{} & \mathsf{Set}_\kappa^{\mathrm{op}}
\end{array}
\tag{7.1.25}
$$

But $A^*\mathsf{Set}_\kappa^{\mathrm{op}} \cong \int_{\mathbb{C}/c} A$ over $\mathbb{C}/c$, and sections of this discrete fibration in Cat correspond uniquely to natural maps $1 \to A$ in $\widehat{\mathbb{C}/c}$. Since 1 is representable in $\widehat{\mathbb{C}/c}$ we can continue (7.1.24) by

$$\dot{\mathcal{V}}_\kappa c \;\cong\; \coprod_{A \in \mathcal{V}_\kappa c} \mathsf{Cat}_{\mathbb{C}/c}(\mathbb{C}/c,\, A^*\mathsf{Set}_\kappa^{\mathrm{op}})$$

$$\cong \coprod_{A \in \mathcal{V}_\kappa c} \widehat{\mathbb{C}/c}(1,\, A)$$

$$\cong \coprod_{A \in \mathcal{V}_\kappa c} A(1_c)$$

$$= \coprod_{A \in \mathcal{V}_\kappa c} \mathsf{El}(\langle c,\, A\rangle)$$

$$= \dot{E}c\,.$$

2. By functoriality of the nerve $\nu : \mathsf{Cat} \to \widehat{\mathbb{C}}$, a sequence of Grothendieck universes

$$\mathsf{Set}_\alpha \subseteq \mathsf{Set}_\beta \subseteq \dots$$

in Set gives rise to a (cumulative) sequence of type-theoretic universes

$$\mathcal{V}_\alpha \rightarrowtail \mathcal{V}_\beta \rightarrowtail \dots$$

in $\widehat{\mathbb{C}}$. More precisely, there is a sequence of cartesian squares,

$$
\begin{array}{ccccc}
\dot{\mathcal{V}}_\alpha & \rightarrowtail & \dot{\mathcal{V}}_\beta & \rightarrowtail & \dots \\
\downarrow & \lrcorner & \downarrow & \lrcorner & \\
\mathcal{V}_\alpha & \rightarrowtail & \mathcal{V}_\beta & \rightarrowtail & \dots ,
\end{array}
\tag{7.1.26}
$$

in the image of $\nu : \mathsf{Cat} \longrightarrow \widehat{\mathbb{C}}$, classifying small maps in $\widehat{\mathbb{C}}$ of increasing size, in the sense of Proposition 7.7.

3. Let $\alpha = 2$ so that $1 \to 2$ is the subobject classifier of Set, and

$$\mathbb{1} = \dot{\mathsf{Set}}_2^{\mathsf{op}} \to \mathsf{Set}_2^{\mathsf{op}} = \mathbb{2}$$

is then a classifier in Cat for *sieves*, i.e. full subcategories $\mathbb{S} \hookrightarrow \mathbb{A}$ closed under the domains of arrows $a \to s$ for $s \in \mathbb{S}$. The nerve $\dot{\mathcal{V}}_2 \to \mathcal{V}_2$ is then the usual subobject classifier $1 \to \Omega$ of $\widehat{\mathbb{C}}$,

$$
\begin{array}{ccccc}
\dot{\mathcal{V}}_2 & = & \nu\mathbb{1} & \xrightarrow{\ \sim\ } & 1 \\
\downarrow & & \downarrow & & \downarrow \\
\mathcal{V}_2 & = & \nu\mathbb{2} & \xrightarrow{\ \sim\ } & \Omega
\end{array}
\tag{7.1.27}
$$

4. For any $X \in \widehat{\mathbb{C}}$, we have an equivalence

$$\widehat{\mathbb{C}}/X \simeq \widehat{\int_{\mathbb{C}} X} \simeq \mathsf{dFib}/_{\int_{\mathbb{C}} X}$$

where, generally, $\mathsf{dFib}/_{\mathbb{D}}$ is the category of discrete fibrations over a category $\mathbb{D}$. This equivalence commutes with composition along discrete fibrations, in the sense that the forgetful functor

$$X_! : \widehat{\mathbb{C}}/X \to \widehat{\mathbb{C}}$$

given by composition along $X \to 1$ agrees (up to canonical isomorphism) with the base change $(p_X)_! \dashv (p_X)^*$ of presheaves along the projection $p_X : \int_{\mathbb{C}} X \to \mathbb{C}$, and with composition along the discrete fibration p_X, as indicated in:

$$\begin{array}{ccccc}
\widehat{\mathbb{C}}/X & \xrightarrow{\ \sim\ } & \widehat{\int_{\mathbb{C}} X} & \xrightarrow{\ \sim\ } & \mathsf{dFib}/_{\int_{\mathbb{C}} X} \\
{\scriptstyle X_!}\downarrow & & {\scriptstyle (p_X)_!}\downarrow & & \downarrow{\scriptstyle p_X\circ(-)} \\
\widehat{\mathbb{C}} & \xrightarrow[\ \sim\ ]{} & \widehat{\mathbb{C}} & \xrightarrow[\ \sim\ ]{} & \mathsf{dFib}/_{\mathbb{C}}.
\end{array}$$

$$\text{(7.1.28)}$$

It follows that the pullback functor $X^* : \widehat{\mathbb{C}} \to \widehat{\mathbb{C}}/X$ commutes with the corresponding right adjoints (one of which is the nerve), and therefore preserves the respective universes,

$$X^* \mathcal{V}_{\mathbb{C}} \cong (p_X)^* v_{\mathbb{C}}(\mathsf{Set}^{\mathsf{op}}) \cong v_{\int_{\mathbb{C}} X}(\mathsf{Set}^{\mathsf{op}}) \cong \mathcal{V}_{\int_{\mathbb{C}} X}.$$

Corollary 7.8 *Let* $\dot{\mathcal{V}}_\alpha \to \mathcal{V}_\alpha$ *classify* α-*small maps in* $\widehat{\mathbb{C}}$, *as in Proposition 7.7. Then for any* $X \in \widehat{\mathbb{C}}$, *the pullback* $X^* \dot{\mathcal{V}}_\alpha \to X^* \mathcal{V}_\alpha$ *classifies* α-*small maps in* $\widehat{\mathbb{C}}/X$.

Remark 7.9 (Standing Assumption Regarding Size) We shall henceforth assume a countable sequence of Grothendieck universes $\mathsf{Set}_\alpha \subset \mathsf{Set}_\beta \subset \ldots \mathsf{Set}_\kappa \subset \ldots$, such that every set occurs in one, with their associated classifiers

$$\mathcal{V}_\alpha \rightarrowtail \mathcal{V}_\beta \rightarrowtail \ldots \mathcal{V}_\kappa \rightarrowtail \ldots$$

in $\widehat{\mathbb{C}}$ for κ-small maps, as in (7.1.26). When constructing further classifiers for (trivial) fibrations from these, as in Propositions 7.11 and 7.17, we similarly assume that these are parametrized by the cardinals κ. Rather than constantly referring to this indexing, however, we shall simply call attention to the relative difference between "small" and "large" where the distinction is relevant. Note that according to this convention, *every map is small* with respect to some universe $\mathcal{V}_\kappa$.

7.2 Classifying Trivial Fibrations

Returning now to the particular presheaf category $\mathsf{cSet} = \mathsf{Set}^{\square^{\mathsf{op}}}$ of cubical sets, recall from Chap. 3 that (uniform) trivial fibration structures on a map $A \to X$ correspond bijectively to relative $+$-algebra structures over X (Definition 3.5). A relative $+$-algebra structure on $A \to X$ is an algebra structure for the pointed polynomial endofunctor $+_X : \mathsf{cSet}/X \longrightarrow \mathsf{cSet}/X$, where recall from (3.1.1),

$$A^+ = \sum_{\varphi:\Phi} A^{[\varphi]} \quad \text{over } X.$$

A +-algebra structure is then a retract $\alpha : A^+ \to A$ over X of the canonical map $\eta_A : A \to A^+$,

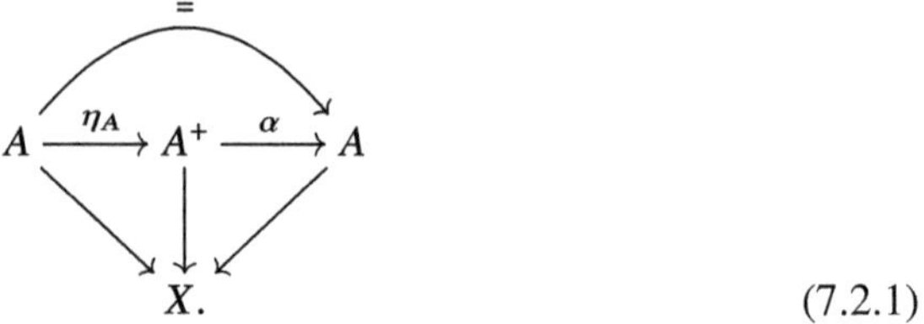

$$(7.2.1)$$

In more detail, let us write $A \to X$ as a family $(A_x)_{x \in X}$, so that $A = \sum_{x:X} A_x \to X$. Since the +-functor acts fiberwise, the object A^+ in (7.2.1) is then the indexing projection

$$\sum_{x:X} A_x^+ \to X.$$

Working in the slice cSet/X, the (relative) exponentials (internal Hom's) $[A^+, A]$ and $[A, A]$ and the "precomposition by η_A" map $[\eta_A, A]$, fit into the following pullback diagram

$$
\begin{array}{ccc}
+\mathsf{Alg}(A) & \longrightarrow & [A^+, A] \\
\downarrow & & \downarrow{\scriptstyle [\eta_A, A]} \\
1 & \xrightarrow{\ \ '\mathrm{id}_A'\ \ } & [A, A].
\end{array}
$$

$$(7.2.2)$$

The constructed object $+\,\mathsf{Alg}(A) \to X$ over X is then the *object of +-algebra structures on* $A \to X$, in the sense that sections $X \to +\,\mathsf{Alg}(A)$ correspond uniquely to +-algebra structures on $A \to X$. Moreover, $+\mathsf{Alg}(A) \to X$ is stable under pullback, in the sense that for any $f : Y \to X$, we have two pullback squares,

$$
\begin{array}{ccc}
f^*A & \longrightarrow & A \\
\downarrow & & \downarrow \\
Y & \xrightarrow{\ \ f\ \ } & X \\
\uparrow & & \uparrow \\
+\mathsf{Alg}(f^*A) & \longrightarrow & +\mathsf{Alg}(A)
\end{array}
$$

$$(7.2.3)$$

because the +-functor, exponentials and pullbacks occurring in the construction of $+\,\mathsf{Alg}(A) \to X$ are themselves all stable.

Let us record what we have learned for future reference, and introduce some new terminology.

Proposition 7.10 *For any map $A \to X$, we have the following* classifying type for trivial fibration structures,

$$\mathsf{TFib}(A) := +\mathsf{Alg}(A) \to X.$$

Sections of $\mathsf{TFib}(A) \to X$ correspond bijectively to $+$-algebra structures on $A \to X$. Moreover, $\mathsf{TFib}(A) \to X$ is stable under pullback, in the sense that for any $f : Y \to X$, we have an iso,

$$\mathsf{TFib}(f^*A) \cong f^*\mathsf{TFib}(A). \tag{7.2.4}$$

It now follows from Proposition 7.7 that, if $A \to X$ is small, then $\mathsf{TFib}(A) \to X$ is itself a pullback of the analogous object $\mathsf{TFib}(\dot{\mathcal{V}}) \to \mathcal{V}$ constructed from the universal small family $\dot{\mathcal{V}} \to \mathcal{V}$ of Proposition 7.7, so there are two pullback squares:

$$\tag{7.2.5}$$

Proposition 7.11 *There is a* universal small trivial fibration

$$\mathsf{T\dot{F}ib} \to \mathsf{TFib}.$$

Every small trivial fibration $A \to X$ is a pullback of $\mathsf{T\dot{F}ib} \to \mathsf{TFib}$ along a canonically determined classifying map $X \to \mathsf{TFib}$.

$$\tag{7.2.6}$$

Proof We can take

$$\mathsf{TFib} := \mathsf{TFib}(\dot{\mathcal{V}}),$$

which comes with its projection $\mathsf{TFib}(\dot{\mathcal{V}}) \to \mathcal{V}$ as in diagram (7.2.5). Then define $p_t : \dot{\mathsf{TFib}} \to \mathsf{TFib}$ by pulling back the universal small family,

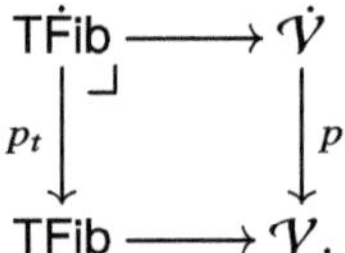

Consider the following diagram, in which all the squares (including the distorted ones) are pullbacks, with the outer one coming from Proposition 7.7 and the lower one from (7.2.5).

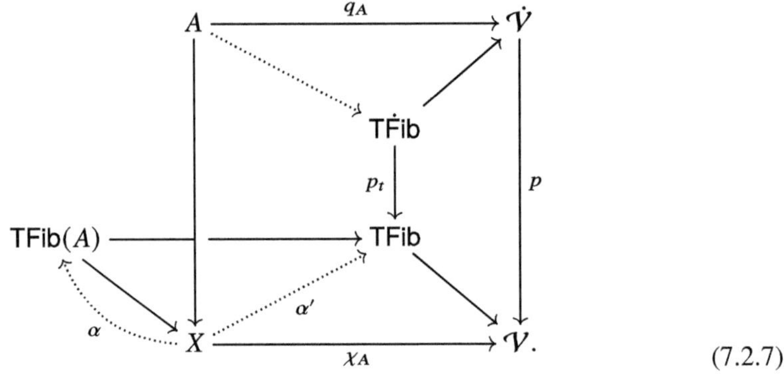

$$(7.2.7)$$

A trivial fibration structure α on $A \to X$ is a section the object of $+$-algebra structures on A, occurring in the diagram as $\mathsf{TFib}(A)$, the pullback of $\mathsf{TFib}(\dot{\mathcal{V}})$ along the classifying map $\chi_A : X \to \mathcal{V}$ for the small family $A \to X$. Such sections correspond uniquely to factorizations α' of χ_A as indicated, which in turn induce pullback squares of the required kind (7.2.6).

Note that the map $p_t : \dot{\mathsf{TFib}} \to \mathsf{TFib}$ has a canonical trivial fibration structure. Indeed, consider the following diagram, in which both squares are pullbacks.

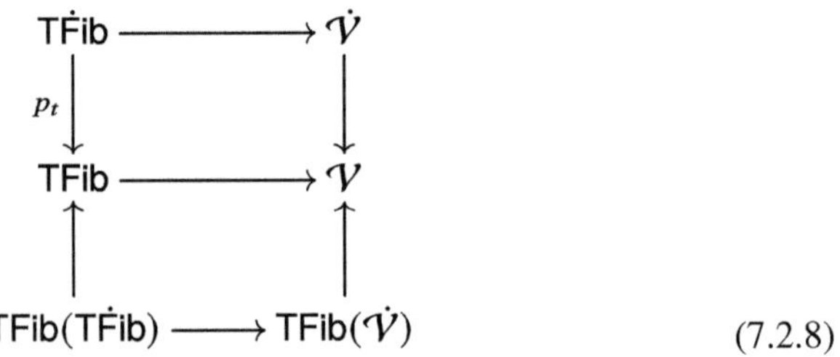

$$(7.2.8)$$

$\mathsf{TFib}(\dot{\mathcal{V}})$ is the object of trivial fibration structures on $\dot{\mathcal{V}} \to \mathcal{V}$, and its pullback $\mathsf{TFib}(\dot{\mathsf{TFib}})$ is therefore the object of trivial fibration structures on $p_t : \dot{\mathsf{TFib}} \to \mathsf{TFib}$. Thus we seek a section of $\mathsf{TFib}(\dot{\mathsf{TFib}}) \to \mathsf{TFib}$. But recall that $\mathsf{TFib} = \mathsf{TFib}(\dot{\mathcal{V}})$ by definition, so the lower pullback square is the pullback of $\mathsf{TFib}(\dot{\mathcal{V}})\to\mathcal{V}$ against itself, which does indeed have a distinguished section, namely the diagonal

$$\Delta : \mathsf{TFib}(\dot{\mathcal{V}})\to\mathsf{TFib}(\dot{\mathcal{V}}) \times_{\mathcal{V}} \mathsf{TFib}(\dot{\mathcal{V}}).$$

$\square$

Since the universal small trivial fibration $\dot{\mathsf{TFib}}\to\mathsf{TFib}$ in cSet from Proposition 7.11 was constructed as $\mathsf{TFib} = \mathsf{TFib}(\dot{\mathcal{V}})$ for the universal small family $\dot{\mathcal{V}}\to\mathcal{V}$, which in turn is stable under pullback by Corollary 7.8, we also have:

Corollary 7.12 *The base change of the universal small trivial fibration*

$$\dot{\mathsf{TFib}}\to\mathsf{TFib}$$

in cSet *along* $\mathrm{I}^* : \mathsf{cSet}\to\mathsf{cSet}/_\mathrm{I}$ *is a universal small trivial fibration in* $\mathsf{cSet}/_\mathrm{I}$.

7.3 Classifying Fibrations

In order to classify fibrations $A\twoheadrightarrow X$, we shall proceed as for trivial fibrations by constructing, for any map $A\to X$, an object $\mathsf{Fib}(A)\to X$ of fibration structures which, moreover, is stable under pullback. We then apply the construction to the universal small family $\dot{\mathcal{V}}\to\mathcal{V}$ of Proposition 7.7 to obtain a universal small fibration. Here we will of course need to distinguish between biased and unbiased fibrations. In Lemma 7.13, we first construct a stable classifying type $\mathsf{Fib}(A)\to X$ for δ-biased fibration structures on any map $A\to X$ in $\mathsf{cSet}/_\mathrm{I}$ where δ is the generic point. In Lemma 7.16 we then transfer the construction along the base change $\mathrm{I}^* : \mathsf{cSet}\to\mathsf{cSet}/_\mathrm{I}$ to obtain a classifier $\mathsf{Fib}(A)\to X$ for unbiased fibration structures on any $A\to X$ in cSet.

The construction of $\mathsf{Fib}(A)\to X$ for biased fibration structures with respect to a point $\delta : 1\to\mathrm{I}$ is already a bit more involved than was that of $\mathsf{TFib}(A)\to X$. In particular, it requires the codomain I of δ to be *tiny*, which is indeed the case for the generic point $\delta : 1\to\mathrm{I}^*\mathrm{I}$ in $\mathsf{cSet}/_\mathrm{I}$ by Lemma 2.7.

7.3.1 *The Classifying Type of Biased Fibration Structures*

A classifying type $\mathsf{Fib}(A)\to X$ of (uniform, δ-biased) fibration structures on a map $p : A\to X$, as defined in Sect. 4.1, can be constructed as follows.

1. First form the pullback-hom $\delta \Rightarrow p : A^\mathrm{I}\to X^\mathrm{I} \times_X A$ with the point $\delta : 1\to\mathrm{I}$, as indicated in the following diagram.

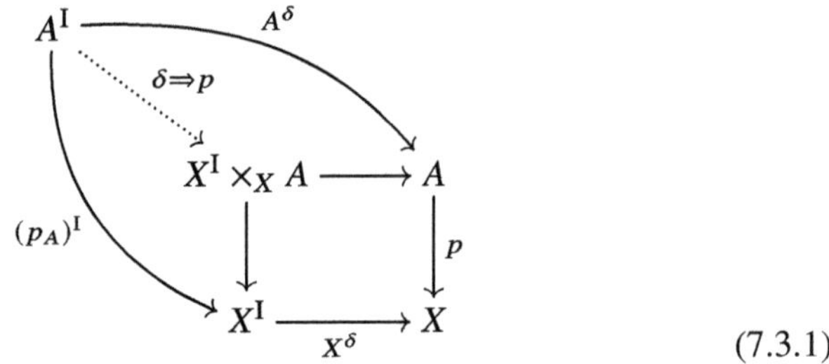

$$(7.3.1)$$

2. A fibration structure on $p : A \to X$ is then a relative $+$-algebra structure on $\delta \Rightarrow p$ in the slice category over its codomain $X^I \times_X A$. To construct a classifier for such structures, let us first relabel the objects and arrows in diagram (7.3.1) as follows:

$$\epsilon := X^\delta : X^I \to X$$

$$A_\epsilon := X^I \times_X A$$

$$\epsilon_A := \delta \Rightarrow p$$

so that the working part of (7.3.1) becomes:

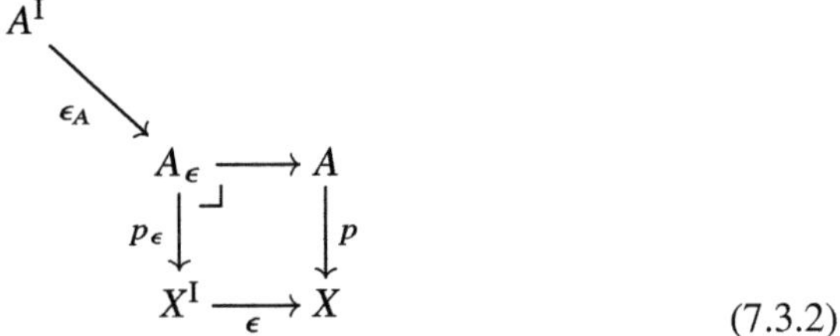

$$(7.3.2)$$

3. Now a relative $+$-algebra structure on ϵ_A (Definition 3.5) is a retract α over A_ϵ of the unit η, as indicated below, where D is simply the domain of the map $(\epsilon_A)^+$ resulting from applying the relative $+$-functor in the slice category over A_ϵ to the object ϵ_A.

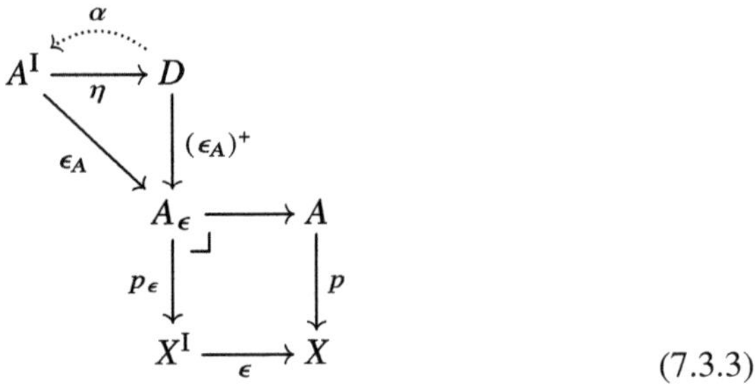

$$(7.3.3)$$

4. As in the construction (7.2.2), there is an object $\mathsf{TFib}(\epsilon_A) = +\mathsf{Alg}(\epsilon_A)$ over A_ϵ of relative $+$-algebra structures on ϵ_A, the sections of which correspond uniquely

to relative +-algebra structures on ϵ_A (and thus to fibration structures on A).

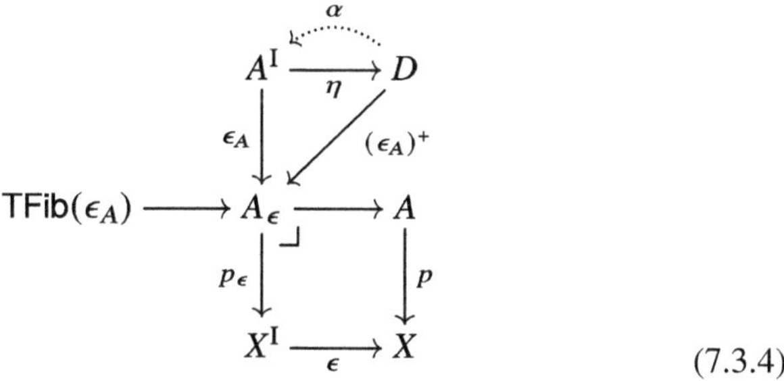

$$(7.3.4)$$

5. Sections of $\mathsf{TFib}(\epsilon_A) \longrightarrow A_\epsilon$ then correspond to sections of its push-forward along p_ϵ, which we shall call F_A:

$$F_A := (p_\epsilon)_* \mathsf{TFib}(\epsilon_A) \,.$$

$$(7.3.5)$$

6. One might now try taking another pushforward of $F_A \to X^{\mathsf{I}}$ along $\epsilon : X^{\mathsf{I}} \to X$ to get the object $\mathsf{Fib}(A) \to X$ that we seek, but unfortunately, this would not be stable under pullback along arbitrary maps $Y \to X$, because the evaluation $\epsilon = X^\delta \cdot X^{\mathsf{I}} \to X$ is not stable in that way. Instead we use the *root* functor, i.e. the right adjoint of the pathspace, $(-)^{\mathsf{I}} \dashv (-)_{\mathsf{I}}$ from Proposition 2.3.

Let $f : F_A \to X^I$ be the map $(p_\epsilon)_* \mathsf{TFib}(\epsilon_A)$ indicated in (7.3.5), and let $\eta : X \to (X^I)_I$ be the unit of the root adjunction at X. Then define $\mathsf{Fib}(A) \to X$ by

$$\mathsf{Fib}(A) := \eta^* f_I$$

as indicated in the following pullback diagram.

$$
\begin{array}{ccc}
\mathsf{Fib}(A) & \longrightarrow & (F_A)_I \\
\downarrow & & \downarrow {\scriptstyle f_I} \\
X & \xrightarrow{\ \eta\ } & (X^I)_I
\end{array}
\qquad (7.3.6)
$$

By adjointness, sections of $\mathsf{Fib}(A) \to X$ then correspond bijectively to sections of $f : F_A \to X^I$.

Lemma 7.13 *For any map $A \to X$ in $\mathsf{cSet}/_I$, the map $\mathsf{Fib}(A) \to X$ in (7.3.6) is a classifying type for δ-biased fibration structures: sections of $\mathsf{Fib}(A) \to X$ correspond bijectively to δ-biased fibration structures on $A \to X$, and the construction is stable under pullback in the sense that for any $f : Y \to X$, we have two pullback squares,*

$$
\begin{array}{ccc}
f^*A & \longrightarrow & A \\
\downarrow & & \downarrow \\
Y & \xrightarrow{\ f\ } & X \\
\uparrow & & \uparrow \\
\mathsf{Fib}(f^*A) & \longrightarrow & \mathsf{Fib}(A)
\end{array}
\qquad (7.3.7)
$$

Proof It is clear from the construction that fibration structures on $A \to X$ correspond bijectively to sections of $\mathsf{Fib}(A) \to X$. We show that $\mathsf{Fib}(A) \to X$ is also stable under pullback. To that end, the relevant steps of the construction are recalled schematically below.

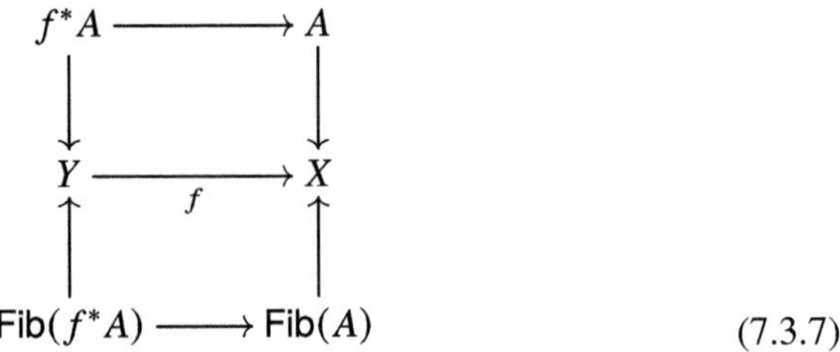

$$(7.3.8)$$

Now consider the following diagram, in which the right hand side consists of the data from (7.3.8), and the front, central square is a pullback.

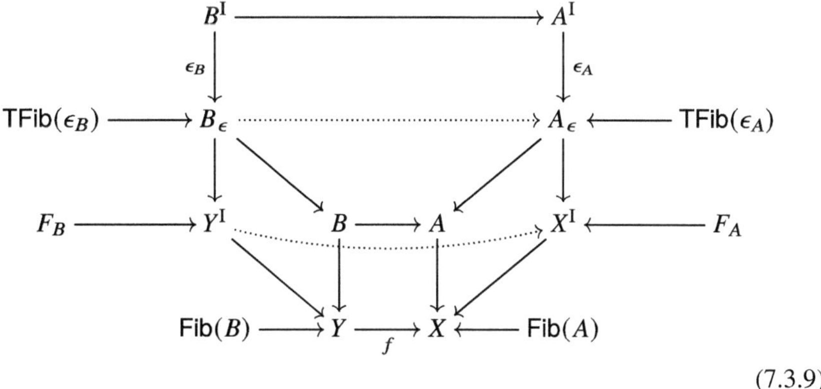

$$(7.3.9)$$

As in the proof of Lemma 6.2, on the left side we repeat the construction with $B \to Y$ in place of $A \to X$. The left face of the indicated (distorted) cube is then also a pullback, whence the back (dotted) face is a pullback, since the two-story square in back is the image of the front pullback square under the right adjoint $(-)^{I}$. Finally, the top rectangle in the back is therefore also a pullback.

It follows that $\mathsf{TFib}(\epsilon_B)$ is a pullback of $\mathsf{TFib}(\epsilon_A)$ along the upper dotted arrow, as in Proposition 7.10, and so the pushforward F_B is a pullback of the corresponding F_A, along the lower dotted arrow (which is f^{I}), by the Beck-Chevalley condition for the dotted pullback square. Let us record this for later reference:

$$F_B \cong (f^{I})^{*} F_A. \tag{7.3.10}$$

It remains to show that $\mathsf{Fib}(B)$ is a pullback of $\mathsf{Fib}(A)$ along $f : Y \to X$, and now it is good that we did not take these to be pushforwards of F_B and F_A, because the floor of the cube need not be a pullback, and so the Beck-Chavalley condition would not apply. Instead, consider the following diagram.

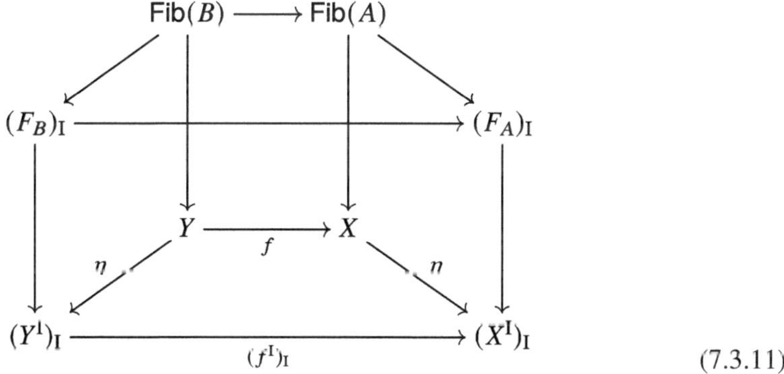

$$(7.3.11)$$

The sides of the cube are pullbacks by the construction of $\mathsf{Fib}(A)$ and $\mathsf{Fib}(B)$. The front face is the root of the pullback (7.3.10) and is thus also a pullback, since the root is a right adjoint. The base commutes by naturality of the unit of the adjunction, and so the back face is also a pullback, as required.   □

Now let us apply the foregoing construction of $\mathsf{Fib}(A)$ to the universal family $\dot{\mathcal{V}}\to\mathcal{V}$ to get $\mathsf{Fib}(\dot{\mathcal{V}})\to\mathcal{V}$, and define the universal small (δ-biased) fibration in $\mathsf{cSet}_{/\mathrm{I}}$ by setting $\mathsf{Fib}:=\mathsf{Fib}(\dot{\mathcal{V}})$ and $\dot{\mathsf{Fib}}\twoheadrightarrow\mathsf{Fib}$ by pulling back the universal family,

$$
\begin{array}{ccc}
\dot{\mathsf{Fib}} & \longrightarrow & \dot{\mathcal{V}} \\
\downarrow & \lrcorner & \downarrow{\scriptstyle p} \\
\mathsf{Fib} & \longrightarrow & \mathcal{V}.
\end{array}
\tag{7.3.12}
$$

The proof of the following then proceeds just as that given for $\mathsf{T\dot{F}ib}\to\mathsf{TFib}$ in Proposition 7.11.

Proposition 7.14 *The map* $\dot{\mathsf{Fib}}\to\mathsf{Fib}$ *constructed in* (7.3.12) *is a* universal small δ-biased *fibration in* $\mathsf{cSet}_{/\mathrm{I}}$: *every small δ-biased fibration* $A\twoheadrightarrow X$ *in* $\mathsf{cSet}_{/\mathrm{I}}$ *is a pullback of* $\dot{\mathsf{Fib}}\twoheadrightarrow\mathsf{Fib}$ *along a canonically determined classifying map* $X\to\mathsf{Fib}$.

$$
\begin{array}{ccc}
A & \longrightarrow & \dot{\mathsf{Fib}} \\
\downarrow & \lrcorner & \downarrow \\
X & \longrightarrow & \mathsf{Fib}
\end{array}
\tag{7.3.13}
$$

Remark 7.15 Proposition 7.14 made no use of the fact that we were working in the slice category $\mathsf{cSet}_{/\mathrm{I}}$ with $\delta : 1\to\mathrm{I}$ the generic point. It holds equally for δ-biased fibrations with respect to any point $\delta : 1\to\mathrm{I}$ of a tiny object I. Thus e.g. it could be used (with obvious adjustment) to construct a classifier for the $\{\delta_0, \delta_1\}$-biased fibrations of Sect. 4.1 in (Cartesian, Dedekind, or other varieties of) cubical sets cSet.

7.3.2 The Classifying Type of Unbiased Fibration Structures

In order to classify *unbiased* fibration structures on maps $A\to X$ in cSet, we first apply the pullback $\mathrm{I}^* : \mathsf{cSet}\to\mathsf{cSet}_{/\mathrm{I}}$ and take the classifier $\mathsf{Fib}(\mathrm{I}^*A)\to\mathrm{I}^*X$ for δ-biased fibration structures, then apply the pushforward $\mathrm{I}_* : \mathsf{cSet}_{/\mathrm{I}}\to\mathsf{cSet}$ and pull the result $\mathrm{I}_*\mathsf{Fib}(\mathrm{I}^*A)\to\mathrm{I}_*\mathrm{I}^*X$ back along the unit $X\to\mathrm{I}_*\mathrm{I}^*X$.

To show that this indeed classifies unbiased fibration structures on $A\to X$, let us first rename the classifying type from Lemma 7.13, which was constructed over I, to $\mathsf{Fib}_i(\mathrm{I}^*A)\to\mathrm{I}^*X$, and then apply I_* to get the map,

$$
\Pi_{i:\mathrm{I}}\mathsf{Fib}_i(\mathrm{I}^*A) := \mathrm{I}_*(\mathsf{Fib}_i(\mathrm{I}^*A))\longrightarrow X^{\mathrm{I}}
$$

in cSet. Then, as just said, we define the desired map $\mathsf{Fib}(A) \to X$ as the pullback along the unit $\rho : X \to X^{\mathrm{I}}$ of $\mathrm{I}^* \dashv \mathrm{I}_*$ as indicated below.

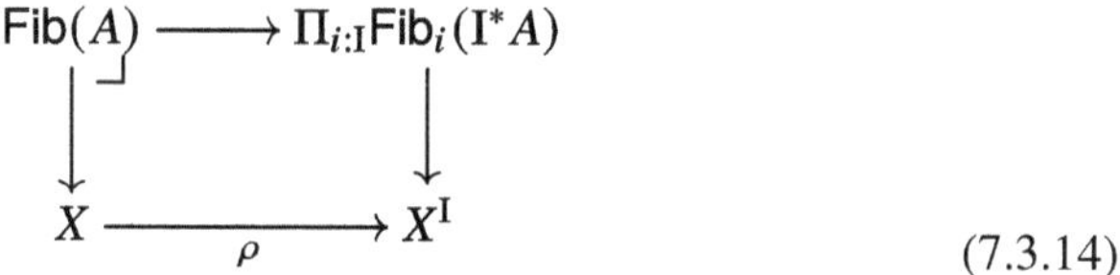

$$
\begin{array}{ccc}
\mathsf{Fib}(A) & \longrightarrow & \Pi_{i:\mathrm{I}}\mathsf{Fib}_i(\mathrm{I}^*A) \\
\downarrow & & \downarrow \\
X & \xrightarrow{\quad\rho\quad} & X^{\mathrm{I}}
\end{array}
\tag{7.3.14}
$$

It now follows immediately from the adjunction $\mathrm{I}^* \dashv \mathrm{I}_*$ that sections of $\mathsf{Fib}(A) \to X$ correspond bijectively to sections of $\mathsf{Fib}_i(\mathrm{I}^*A) \to \mathrm{I}^*X$ over I, and thus to *unbiased fibration structures* on $A \to X$.

Lemma 7.16 *For any map $A \to X$ in cSet, the map $\mathsf{Fib}(A) \to X$ in (7.3.14) is a classifying type for unbiased fibration structures: sections of $\mathsf{Fib}(A) \to X$ correspond bijectively to unbiased fibration structures on $A \to X$, and the construction is stable under pullback in the expected sense (as in Lemma 7.13).*

Proof It remains only to check the stability, but since both of the adjoints in $\mathrm{I}^* \dashv \mathrm{I}_* : \mathsf{cSet}_{/\mathrm{I}} \to \mathsf{cSet}$ preserve pullbacks, this follows easily from the fact that the classifying types Fib_i are stable under pullback by Lemma 7.13. $\square$

Finally, we can again take $\mathsf{Fib} := \mathsf{Fib}(\dot{\mathcal{V}})$ to now obtain a universal small *unbiased* fibration $\dot{\mathsf{Fib}} \to \mathsf{Fib}$ in cSet, as in (7.3.12), and the proof can conclude just as in that for Proposition 7.11.

Proposition 7.17 *The map $\dot{\mathsf{Fib}} \to \mathsf{Fib}$ just constructed is a universal small unbiased fibration in cSet: every small unbiased fibration $A \twoheadrightarrow X$ is a pullback of $\dot{\mathsf{Fib}} \twoheadrightarrow \mathsf{Fib}$ along a canonically determined classifying map $X \to \mathsf{Fib}$.*

$$
\begin{array}{ccc}
A & \longrightarrow & \dot{\mathsf{Fib}} \\
\downarrow & & \downarrow \\
X & \longrightarrow & \mathsf{Fib}
\end{array}
\tag{7.3.15}
$$

Remark 7.18 Recall from Proposition 7.8 that the universe in the slice category $\mathsf{cSet}_{/\mathrm{I}}$ is the pullback of the universe $\mathcal{V}$ from cSet along the base change $\mathrm{I}^* : \mathsf{cSet} \to \mathsf{cSet}_{/\mathrm{I}}$. Thus in the construction just given of the classifier $\dot{\mathsf{Fib}} \to \mathsf{Fib}$ for unbiased fibrations in cSet we are first building the classifying type

$$\mathsf{Fib}_i(\mathrm{I}^*\dot{\mathcal{V}})\to\mathrm{I}^*\mathcal{V}$$

for δ-biased fibration structures on the universal family in $\mathsf{cSet}_{/\mathrm{I}}$, and then taking a pushforward $\mathrm{I}_* : \mathsf{cSet}_{/\mathrm{I}}\to\mathsf{cSet}$ to obtain the (base of the) classifier for unbiased fibrations as the pullback along the unit:

$$
\begin{array}{ccc}
\mathsf{Fib}(\dot{\mathcal{V}}) & \longrightarrow & \Pi_{i:\mathrm{I}}\mathsf{Fib}_i(\mathrm{I}^*\dot{\mathcal{V}}) \\
\downarrow & & \downarrow \\
\mathcal{V} & \xrightarrow{\ \rho\ } & \mathcal{V}^{\mathrm{I}}
\end{array}
\tag{7.3.16}
$$

We remark for later reference that this classifying type $\mathsf{Fib} = \mathsf{Fib}(\dot{\mathcal{V}})\to\mathcal{V}$ for unbiased fibration structures can therefore be constructed as the pushforward of the classifier $\mathsf{Fib}_i(\mathrm{I}^*\dot{\mathcal{V}})\to\mathrm{I}^*\mathcal{V}$ for δ-biased fibration structures along the projection $q : \mathrm{I}^*\mathcal{V} = \mathrm{I} \times \mathcal{V}\to\mathcal{V}$ indicated below.

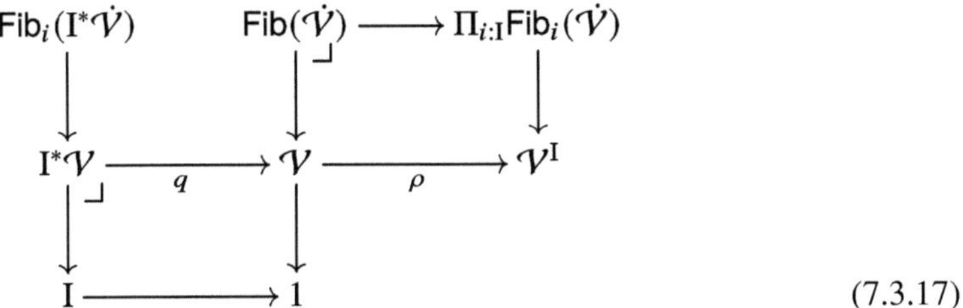

$$
\tag{7.3.17}
$$

We record this fact as:

Corollary 7.19 $\mathsf{Fib} = \Sigma_{\mathcal{V}}\, q_*\mathsf{Fib}_i(\mathrm{I}^*\dot{\mathcal{V}})$.

The reader may also find it illuminating to reconsider the construction of the universal small unbiased fibration in more type theoretic terms. It was defined to be $\dot{\mathsf{Fib}}\twoheadrightarrow\mathsf{Fib} = \mathsf{Fib}(\dot{\mathcal{V}})$, for the universal family $\dot{\mathcal{V}}\to\mathcal{V}$, with $\dot{\mathsf{Fib}}$ the pullback of $\dot{\mathcal{V}}\to\mathcal{V}$ along the canonical projection $\mathsf{Fib}(\dot{\mathcal{V}})\to\mathcal{V}$. Since, type theoretically, we have $\dot{\mathcal{V}} = \Sigma_{A:\mathcal{V}}A$, by the stability of the classifying type $\mathsf{Fib}(-)$ we can write $\mathsf{Fib} = \Sigma_{A:\mathcal{V}}\mathsf{Fib}(A)$ so that:

$$\dot{\mathsf{Fib}} = \Sigma_{A:\mathcal{V}}\mathsf{Fib}(A) \times A \longrightarrow \Sigma_{A:\mathcal{V}}\mathsf{Fib}(A) = \mathsf{Fib}.$$

7.4   Realignment for Fibration Structure

The realignment for families of Proposition 7.6 will need to be extended to (structured) fibrations. Our approach makes use of the notion of a *weak proposition*.

Informally, a map $P \to X$ may be said to be a weak proposition if it is "conditionally contractible", in the sense that it is contractible if it has a section (recall that a proposition may be defined internally as a type that is "contractible if inhabited"). More formally, we have the following.

Definition 7.20 A map $P \to X$ is said to be a *weak proposition* if the projection $P \times_X P \to P$ is a trivial fibration.

$$
\begin{array}{ccc}
P^2 & \longrightarrow & P \\
{\scriptstyle\sim}\downarrow & \lrcorner & \downarrow \\
P & \longrightarrow & X.
\end{array}
\tag{7.4.1}
$$

Note that if either projection is a trivial fibration, then both are.

As an object over the base, a weak proposition is thus one that "thinks it is contractible". The key fact needed for realignment is the following.

Lemma 7.21 *For any $A \to X$, the classifying type $\mathsf{TFib}(A) \to X$ is a weak proposition. Moreover, the same is true for $\mathsf{Fib}(A) \to X$ (both the biased and unbiased versions) if the cofibrations are closed under exponentiation by the interval I.*

Proof Let $A \to X$ and consider the following diagram, in which we have written $A' = \mathsf{TFib}(A) \times_X A$ and $\mathsf{TFib}(A)^2 = \mathsf{TFib}(A) \times_X \mathsf{TFib}(A)$.

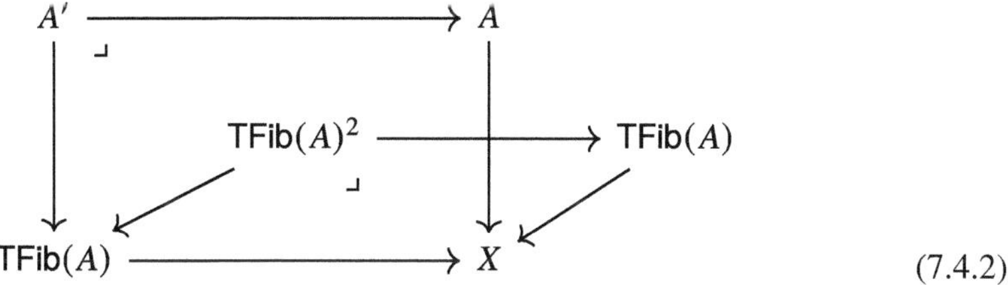

$$
\tag{7.4.2}
$$

Since TFib is stable under pullback (by Proposition 7.10), we have $\mathsf{TFib}(A)^2 \cong \mathsf{TFib}(A')$, and since $\mathsf{TFib}(A)^2$ has a canonical section, $A' \to \mathsf{TFib}(A)$ is therefore a trivial fibration. Inspecting the definition of $\mathsf{TFib}(A) = +\mathsf{Alg}(A)$ in (7.2.2), we see that if a map $A \to X$ is a trivial fibration, then so is $\mathsf{TFib}(A) \to X$ (since $\eta : A \to A^+$ is always a cofibration). Thus $\mathsf{TFib}(A)^2 \cong \mathsf{TFib}(A') \to \mathsf{TFib}(A)$ is also a trivial fibration.

For $\mathsf{Fib}(A) \to X$, with reference to the construction (7.3.8) we use the foregoing to infer that $\mathsf{TFib}(\epsilon_A) \to A_\epsilon$ is a weak proposition, and so therefore is its pushforward $F_A = (p_c)_* \mathsf{TFib}(\epsilon_A) \to X^\mathrm{I}$ along the projection $p_c : A_c = X^\mathrm{I} \times_X A \to X^\mathrm{I}$, since pushforward clearly preserves weak propositions. Applying the root $(-)_\mathrm{I}$ preserves trivial fibrations, by the assumption that its left adjoint $(-)^\mathrm{I}$ preserves cofibrations, and so, as a right adjoint, it also preserves weak propositions. Therefore $(F_A)_\mathrm{I} \to (X^\mathrm{I})_\mathrm{I}$ is a weak proposition, but then so is its pullback along the unit $X \to (X^\mathrm{I})_\mathrm{I}$, which is $\mathsf{Fib}_i(A) \to X$, the classifier for δ-biased fibration structures. The same reasoning shows that $\mathsf{Fib}(A) = \rho^* \Pi_{i:\mathrm{I}} \mathsf{Fib}_i(\mathrm{I}^* A)$ (as in (7.3.14)) is also a weak proposition.   □

In light of Lemma 7.21 we shall henceforth assume the following, as a final axiom on cofibrations:

(C8) The pathobject functor preserves cofibrations: thus $c : A \rightarrowtail B$ implies that $c^{\mathrm{I}} : A^{\mathrm{I}} \rightarrowtail B^{\mathrm{I}}$.

Now, by Propositions 7.14 and 7.17 we have universal small δ-biased and unbiased fibrations, the former in $\mathsf{cSet}/_{\mathrm{I}}$, the latter in cSet. The following remarks apply to both, which we refer to neutrally as $\dot{\mathcal{U}} \twoheadrightarrow \mathcal{U}$. The base object $\mathcal{U}$ is (the domain of) the classifying type $\mathsf{Fib}(\dot{\mathcal{V}}) \to \mathcal{V}$, where $\dot{\mathcal{V}} \to \mathcal{V}$ is the universal small family. Type theoretically, this object can be written as

$$\mathcal{U} = \Sigma_{E:\mathcal{V}} \mathsf{Fib}(E) \,,$$

which comes with the canonical projection

$$\mathcal{U} = \Sigma_{E:\mathcal{V}} \mathsf{Fib}(E) \longrightarrow \mathcal{V} \,.$$

In these terms, a fibration $E \twoheadrightarrow X$ is a pair $\langle E, e \rangle$, consisting of the underlying family $E \to X$, equipped with a fibration structure $e : \mathsf{Fib}(E)$. Lemma 7.21 then allows us to establish the following, which was first isolated in [67] (as condition (2'), also see [38]). It holds for both biased and unbiased fibrations, and will be used in the sequel to "correct" the fibration structure on certain maps.

Lemma 7.22 (Realignment for Fibrations)   *Given a fibration $F \twoheadrightarrow X$ and a cofibration $c : C \rightarrowtail X$, let $f_c : C \to \mathcal{U}$ classify the pullback $c^* F \twoheadrightarrow C$. Then there is a classifying map $f : X \to \mathcal{U}$ for F with $f \circ c = f_c$.*

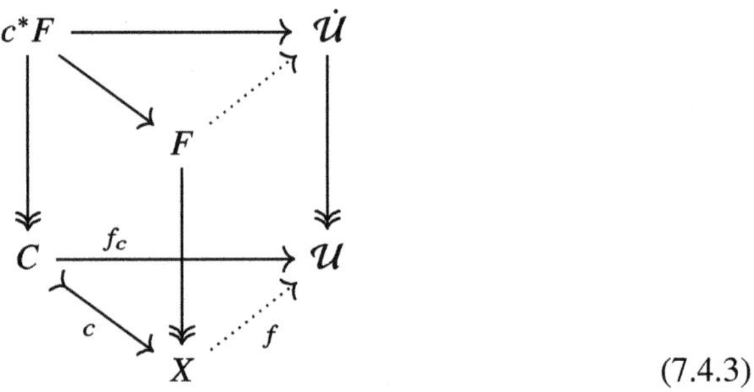

$$(7.4.3)$$

Proof First, let $|f_c| : C \to \mathcal{V}$ be the composite of $f_c : C \to \mathcal{U}$ with the canonical projection $\mathcal{U} \to \mathcal{V}$, thus classifying the underlying family $c^* F \to C$. Next, let $f_0 : X \to \mathcal{V}$ classify the underlying family $F \to X$. We may assume that $f_0 \circ c = |f_c|$ by realignment for families, Proposition 7.6.

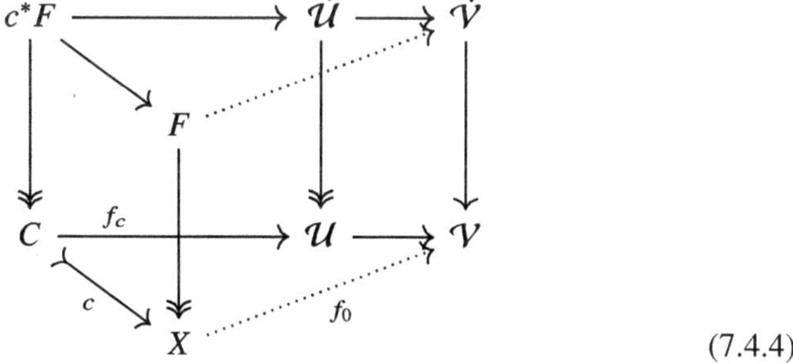

$$(7.4.4)$$

Since $F \twoheadrightarrow X$ is a fibration, there is a lift $f_1 : X \to \mathcal{U}$ of f_0 classifying the fibration structure. We thus have the following commutative diagram in the base of (7.4.4).

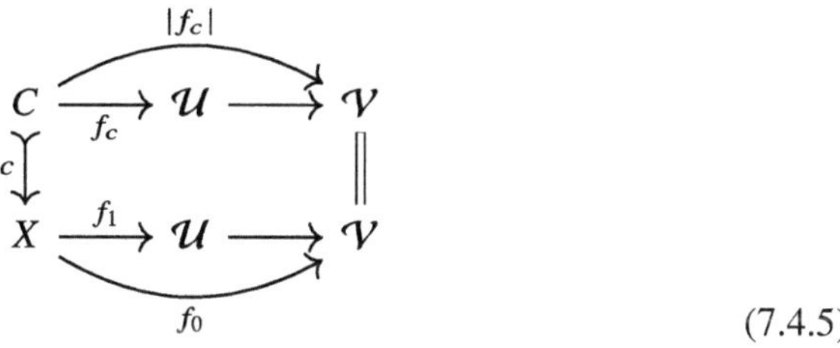

$$(7.4.5)$$

Now pull $\mathcal{U} \to \mathcal{V}$ back against itself and rearrange the previous data to give (the solid part of) the following, which also commutes.

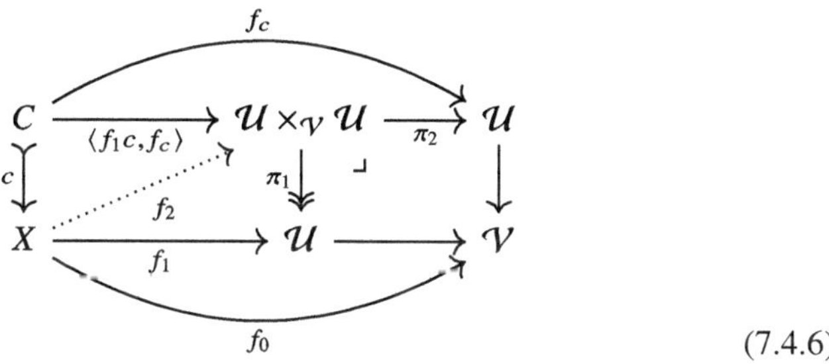

$$(7.4.6)$$

Since $\mathcal{U} = \mathsf{Fib}(\dot{\mathcal{V}}) \to \mathcal{V}$ is a weak proposition by Lemma 7.21 and (C8), the projection $\pi_1 : \mathcal{U} \times_{\mathcal{V}} \mathcal{U} \twoheadrightarrow \mathcal{U}$ is a trivial fibration, so there is a diagonal filler $f_2 : X \to \mathcal{U} \times_{\mathcal{V}} \mathcal{U}$ as indicated. Taking $f := \pi_2 \circ f_2 : X \to \mathcal{U} \times_{\mathcal{V}} \mathcal{U} \to \mathcal{U}$ gives another classifying map for the fibration structure on $F \to X$, for which $f \circ c = f_c$ as required. $\qquad\square$

Chapter 8
The Equivalence Extension Property

We shall define the equivalence extension property in the category cSet of cubical sets, which is closely related to the *univalence* of the universal fibration $\dot{\mathcal{U}} \twoheadrightarrow \mathcal{U}$ constructed in Sect. 7.3 (see [67]). It will be used in Chap. 9 to show that the base object $\mathcal{U}$ is fibrant. The proof of the equivalence extension property given here is a reformulation of a type-theoretic argument due to Coquand, cf. [28], which in turn is a modification of the original argument of Voevodsky, which can be found in [50]. See [66] for another reformulation.

8.1 The Sliced Premodel Structure

We begin by recalling some basic facts and making some simple observations that are well-known in general model categories, but need to be checked again here, because we do not yet have a full model structure. The reader is reminded that the word "fibration" unqualified always refers to *unbiased* fibrations as in Definition 4.6. First, for any object $Z \in \mathsf{cSet}$, the slice category $\mathsf{cSet}/_Z$ inherits the premodel structure of Proposition 5.4 from cSet via the forgetful functor

$$Z_! : \mathsf{cSet}/_Z \longrightarrow \mathsf{cSet}\,.$$

In more detail:

Definition 8.1 A map $f : X \to Y$ over Z is a *(trivial) cofibration or (trivial) fibration* over Z just if it is one in cSet after forgetting the Z-indexing via $Z_! : \mathsf{cSet}/_Z \to \mathsf{cSet}$. This will be called the *(relative or) sliced premodel structure* on $\mathsf{cSet}/_Z$. Accordingly, a map $f : X \to Y$ over Z will be called a *weak equivalence* over Z just if it factors over Z as a trivial fibration over Z after a trivial cofibration over Z, which therefore holds just if it is a weak equivalence in cSet after forgetting the Z-indexing.

© The Author(s), under exclusive license to Springer Nature Switzerland AG 2026

S. Awodey, *Cartesian Cubical Model Categories*, Lecture Notes in Mathematics 2385, https://doi.org/10.1007/978-3-032-08730-0_8

That the specification in Definition 8.1 actually does determine a premodel structure is a consequence of Proposition 5.4, and the well-known fact that (pre-)model structures are stable under slicing in this way, cf. [41]. In more detail:

Lemma 8.2 *A map $f : X \to Y$ over Z is a fibration (respectively, a trivial fibration) over Z if, and only if, it lifts on the right in the slice category $\mathsf{cSet}/_Z$ against all trivial cofibrations (respectively, cofibrations) over Z.*

Proof Let $X \xrightarrow{f} Y \xrightarrow{p_Y} Z$, regarded as a map in the slice category over Z, with $p_X = p_Y \circ f : X \to Z$. Then by definition f is a fibration in $\mathsf{cSet}/_Z$ just if $f : X \to Y$ is a fibration in the total category cSet, which holds just if f lifts on the right against all trivial cofibrations $t : A \rightarrowtail B$ in cSet. But every lifting problem of the form $t \pitchfork f$ in cSet,

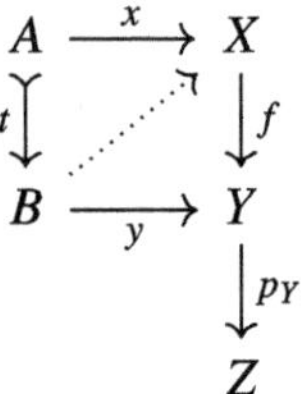

gives rise to a corresponding one over Z, just by composing everything with $p_Y : Y \to Z$. Moreover, the evident resulting map $A \xrightarrow{t} B \to Z$ is then a trivial cofibration over Z, and every such lifting problem for f over Z arises in this way. Finally, the diagonal fillers for the resulting lifting problem in $\mathsf{cSet}/_Z$ are exactly the diagonal fillers for the original one in cSet. Thus the map f over Z is a fibration over Z just in case it lifts on the right over Z against all trivial cofibrations over Z, as claimed. The case of trivial fibrations and cofibrations is exactly analogous. □

Lemma 8.3 *A map $f : X \to Y$ over Z is a cofibration (respectively, a trivial cofibration) over Z if, and only if, it lifts on the left in the slice category $\mathsf{cSet}/_Z$ against all trivial fibrations (respectively, fibrations) over Z.*

Proof Let $X \xrightarrow{f} Y \xrightarrow{p_Y} Z$, regarded as a map in the slice category over Z, with $p_X = p_Y \circ f : X \to Z$. Then by definition f is a cofibration in $\mathsf{cSet}/_Z$ just if $Z_! f : Z_! X \to Z_! Y$ is a cofibration in the total category cSet, which holds just if $Z_! f$ lifts on the left against all trivial fibrations $t : E \twoheadrightarrow F$ in cSet. But every lifting problem of the form $Z_! f \pitchfork t$ in cSet,

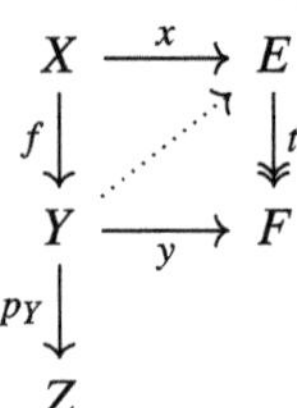

gives rise to a corresponding one over Z of the form $f \pitchfork Z^*t$, by pulling t back along $Z \to 1$. Moreover, since trivial fibrations are stable under pullback in cSet, the pullback Z^*t is a trivial fibration, and so Z^*t is a trivial fibration over Z. Thus f is a cofibration in cSet/Z if and only if $f \pitchfork Z^*t$ in cSet/Z for all trivial fibrations $t : E \twoheadrightarrow F$ in cSet.

Now observe that for any map $A \xrightarrow{g} B \xrightarrow{p_B} Z$ over Z, with $p_A = p_B \circ g$, the following unit square is a pullback, as indicated below,

$$
\begin{array}{ccccc}
A & \xrightarrow{\;\eta_A\;} & Z^*Z_! A & =\!=\!= & Z \times A \\
{\scriptstyle g}\downarrow & \lrcorner & \downarrow{\scriptstyle Z^*Z_! g} & & \downarrow{\scriptstyle Z \times g} \\
B & \xrightarrow[\;\eta_B\;]{} & Z^*Z_! B & =\!=\!= & Z \times B \\
\downarrow & & \downarrow & & \downarrow \\
Z & =\!=\!= & Z & =\!=\!= & Z,
\end{array}
\tag{8.1.1}
$$

because the graph $\eta_A = \langle p_A, 1_Z \rangle : A \to Z \times A$ is a pullback of $\Delta_Z = \langle 1_Z, 1_Z \rangle : Z \to Z \times Z$ along $1_Z \times p_A : Z \times A \to Z \times Z$, and similarly for η_B. Thus in particular, every trivial fibration $A \xrightarrow{g} B \xrightarrow{p_B} Z$ over Z is a pullback over Z of one of the form $Z^*g : Z^*A \twoheadrightarrow Z^*B$ for a trivial fibration $g : A \twoheadrightarrow B$ in cSet. Therefore f is a cofibration in cSet/Z if and only if $f \pitchfork g$ in cSet/Z for all trivial fibrations $g : A \twoheadrightarrow B$ in cSet/Z, as claimed. The case of trivial cofibrations and fibrations is exactly analogous. $\qquad\square$

Since factoring a map in the slice category is evidently given simply by factoring it after forgetting the indexing, we now have:

Proposition 8.4 *The specification in Definition 8.1 determines a premodel structure on cSet/Z for any object $Z \in \mathsf{cSet}$.*

The reader is warned that when $Z = \mathrm{I}$ there is a possibility of confusion with the δ-biased fibrations in $\mathsf{cSet}/_\mathrm{I}$, which do not in general agree with the I-sliced (unbiased) fibrations.

In order to verify the axioms (C1)–(C8) for cofibrations, let $Z^*1 \rightrightarrows Z^*\mathrm{I}$ in cSet/Z be the result of pulling the interval $1 \rightrightarrows \mathrm{I}$ back along $Z \to 1$, to obtain a bipointed object in cSet/Z that we shall write as,

$$
\delta_0, \delta_1 : 1_Z \rightrightarrows \mathrm{I}_Z .
\tag{8.1.2}
$$

Observe that $1_Z + 1_Z \cong Z^*1 + Z^*1 \rightarrowtail Z^*\mathrm{I}$ since the pullback functor $Z^* : \mathsf{cSet} \to \mathsf{cSet}/Z$ preserves (co)limits and cofibrations.

Proposition 8.5 *Taking $\delta_0, \delta_1 : 1_Z \rightrightarrows \mathrm{I}_Z$ as an interval, the axioms (C1)–(C8) for cofibrations are satisfied in cSet/Z*

Proof The (relative) cofibration classifier in $\mathsf{cSet}/_Z$ is the pullback Z^*t : $Z^*1 \rightarrowtail Z^*\Phi$, which we shall write as

$$t_Z : 1_Z \rightarrowtail \Phi_Z . \tag{8.1.3}$$

For axiom (C8), observe that for a map $c : A \to B$ in $\mathsf{cSet}/_Z$, the exponential c^{I_Z} : $A^{\mathrm{I}_Z} \to A^{\mathrm{I}_Z}$ in $\mathsf{cSet}/_Z$ fits into a unit pullback square of the form (8.1.1),

$$\begin{array}{ccccc}
A^{\mathrm{I}_Z} & \xrightarrow{\ \eta_A\ } & Z^*Z_!(A^{\mathrm{I}_Z}) & \xrightarrow{\ \cong\ } & Z^*\big(Z_!(A)^{\mathrm{I}}\big) \\
{\scriptstyle c^{\mathrm{I}_Z}}\Big\downarrow & \lrcorner & \Big\downarrow{\scriptstyle Z^*Z_!(c^{\mathrm{I}_Z})} & & \Big\downarrow{\scriptstyle Z^*\big(Z_!(c)^{\mathrm{I}}\big)} \\
B^{\mathrm{I}_Z} & \xrightarrow[\ \eta_B\ ]{} & Z^*Z_!(B^{\mathrm{I}_Z}) & \xrightarrow[\ \cong\ ]{} & Z^*\big(Z_!(B)^{\mathrm{I}}\big)
\end{array} \tag{8.1.4}$$

So if c is a cofibration, so is c^{I_Z}. The other axioms are routine and left to the reader.
$\qquad\square$

Lemma 8.6 *For any cubical set Z, we have the following relative versions of the pushout-product and pullback-hom conditions involving the interval in the slice category $\mathsf{cSet}/_Z$.*

1. *If $c : A \rightarrowtail B$ is a cofibration in $\mathsf{cSet}/_Z$, then the pushout-product formed in $\mathsf{cSet}/_Z$ with $\delta_0 : 1_Z \rightarrowtail \mathrm{I}_Z$, written*

$$c \otimes_Z \delta_0 : B +_A (A \times_Z \mathrm{I}_Z) \longrightarrow B \times_Z \mathrm{I}_Z ,$$

   is a trivial cofibration (and similarly for $\delta_1 : 1_Z \rightarrowtail \mathrm{I}_Z$).
2. *If $f : X \twoheadrightarrow Y$ is a fibration in $\mathsf{cSet}/_Z$, then the pullback-hom formed in $\mathsf{cSet}/_Z$ with $\delta_0 : 1_Z \rightarrowtail \mathrm{I}_Z$, written*

$$\delta_0 \Rightarrow_Z f : X^{\mathrm{I}_Z} \longrightarrow Y^{\mathrm{I}_Z} \times_Z X ,$$

   is a trivial fibration (and similarly for $\delta_1 : 1_Z \rightarrowtail \mathrm{I}_Z$).

Proof For (1), the pushout-product $c \otimes_Z \delta_0 : D \to B \times_Z \mathrm{I}_Z$ over Z is equal to the (non-relative) pushout-product $c \otimes \delta_0 : D \to B \times \mathrm{I}$, because $Z^*\delta_0 : Z^*1 \to Z^*\mathrm{I}$ is constant over Z, so

$$B \times_Z \mathrm{I}_Z \cong B \times \mathrm{I} ,$$

and similarly for A (and pushouts in the slice are created by the forgetful functor $\mathsf{cSet}/_Z \to \mathsf{cSet}$). Thus, briefly,

$$c \otimes_Z Z^*\delta_0 = c \otimes \delta_0 ,$$

which is indeed a trivial cofibration.

(2) follows from (1) and Lemma 8.3, together with the usual adjunction between $\otimes_Z$ and $\Rightarrow_Z$. $\square$

In order to apply the results on weak equivalences from Chap. 5 in arbitrary slice categories $\mathsf{cSet}/_Z$ we shall also require the notions of homotopy equivalence over Z and weak homotopy equivalence over Z. We first use the relative interval $\delta_0, \delta_1 : 1_Z \rightrightarrows I_Z$ (8.1.2) to define homotopy between maps over Z in the expected way, namely:

Definition 8.7 For any object Z and maps $f, g : X \rightrightarrows Y$ in $\mathsf{cSet}/_Z$, a *homotopy over* Z, written

$$\vartheta : f \sim_Z g,$$

is a map over Z,

$$\vartheta : I_Z \times_Z X \longrightarrow Y,$$

such that $\vartheta \circ \iota_0 = f$ and $\vartheta \circ \iota_1 = g$,

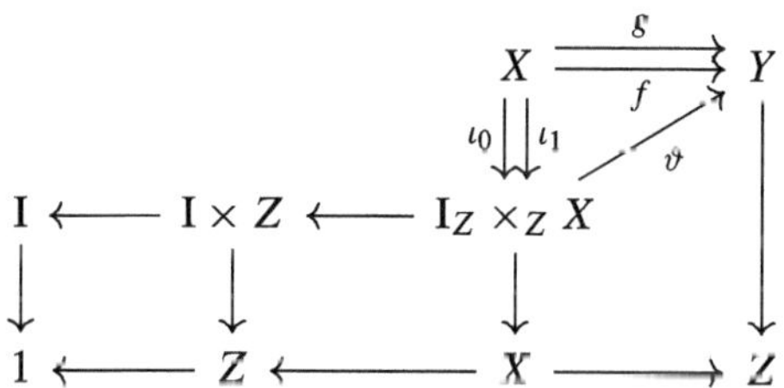

$$(8.1.5)$$

where, as usual, ι_0, ι_1 are the canonical inclusions into the ends of the cylinder,

$$\iota_\epsilon : X \cong 1_Z \times_Z X \xrightarrow{\ \delta_\epsilon \times_Z X\ } I_Z \times_Z X, \qquad \epsilon = 0, 1.$$

Lemma 8.8 *For any object Z and maps $f, g : X \rightrightarrows Y$ in $\mathsf{cSet}/_Z$, a homotopy over Z determines a homotopy of the underlying maps by applying the functor $Z_! : \mathsf{cSet}/_Z \to \mathsf{cSet}$ that forgets the Z-indexing,*

$$\vartheta : f \sim_Z g \quad \mapsto \quad Z_!\vartheta : Z_!f \sim Z_!g.$$

Proof Consider the following diagram depicting a homotopy $\vartheta : f \sim_Z g$ over Z.

Since the lower left two squares are pullbacks, we have $I_Z \times_Z X \cong I \times X$. So applying $Z_!$ to ϑ results in a homotopy $Z_!\vartheta : Z_!f \sim Z_!g$.

Note that an arbitrary homotopy $\varphi : f \sim g$ will not result in one over Z, however, unless φ commutes with the indexing maps to Z.　　　　　　　　　$\square$

Proposition 8.9 *For any object Z, the relation of homotopy over Z between maps $f, g : X \rightrightarrows Y$ over Z is preserved by pre- and post-composition. If $X \twoheadrightarrow Z$ and $Y \twoheadrightarrow Z$ are both fibrations, then the relation $f \sim_Z g$ of maps between them is an equivalence relation.*

The proof is essentially the same as the corresponding one for homotopy over 1, Proposition 5.6, with the exception that both X and Y are required to be fibrant objects over Z, so that the exponential Y^X, taken over Z, is also a fibration $Y^X \twoheadrightarrow Z$ (by Corollary 6.7).

Next we define a *connected components* functor on the full subcategory $\mathsf{Fib}_Z \hookrightarrow \mathsf{cSet}/_Z$ of fibrations over Z,

$$(\pi_0)_Z : \mathsf{Fib}_Z \rightarrow \mathsf{Set},$$

by taking the global sections of a fibration $F \twoheadrightarrow Z$, modulo the relation $\sim_Z$ of homotopy over Z. In more detail, for $F \twoheadrightarrow Z$ in Fib_Z let $(\pi_0)_Z(F)$ be the coequalizer,

$$\mathrm{Hom}_Z(I_Z, F) \rightrightarrows \mathrm{Hom}_Z(1_Z, F) \rightarrow (\pi_0)_Z(F), \tag{8.1.6}$$

where the two maps are given by precomposition with the interval $1_Z \rightrightarrows I_Z$ over Z, and the Hom-sets are those in $\mathsf{cSet}/_Z$.

For fibrations $X \twoheadrightarrow Z$ and $F \twoheadrightarrow Z$ we then again have

$$(\pi_0)_Z(F^X) = \mathrm{Hom}_Z(X, F)/\sim_Z,$$

so $(\pi_0)_Z(F^X)$ is the set $[X, F]_Z$ of Z-homotopy equivalence classes of maps $X \rightarrow F$ over Z. For maps over a base object Z, we can then define the notions of homotopy equivalence over Z and, between fibrations, weak homotopy equivalence over Z as before (cf. Chap. 5):

Definition 8.10 Let Z be any object in cSet, and let $X \rightarrow Z$ and $Y \rightarrow Z$, regarded as objects over Z.

1. A map $f : X \rightarrow Y$ over Z is a *homotopy equivalence over Z* if there is a map $g : Y \rightarrow X$ over Z and two homotopies over Z,

$$\vartheta : g \circ f \sim_Z 1_X, \qquad \varphi : f \circ g \sim_Z 1_Y.$$

2. For $X \twoheadrightarrow Z$ and $Y \twoheadrightarrow Z$ fibrations, a map $f : X \rightarrow Y$ over Z is a *weak homotopy equivalence over Z* if for every fibration $F \twoheadrightarrow Z$, the precomposition map over Z,

$$F^f : F^Y \rightarrow F^X,$$

is bijective on connected components,

$$(\pi_0)_Z(F^f) : (\pi_0)_Z(F^Y) \cong (\pi_0)_Z(F^X),$$

where the indicated exponentials are taken in the slice category over Z.

The proof of the following is analogous to that of the corresponding facts for the case $Z = 1$ (Lemmas 5.8 and 5.14).

Lemma 8.11 *For any object Z, the homotopy equivalences over Z between fibrations $F \twoheadrightarrow Z$ satisfy the 3-for-2 condition, as do the weak homotopy equivalences over Z.*

Proposition 8.12 *For any object Z and fibrations $X \twoheadrightarrow Z$ and $Y \twoheadrightarrow Z$, the following conditions are equivalent for any map $f : X \to Y$ over Z.*

1. *$f : X \to Y$ is a weak equivalence over Z,*
2. *$f : X \to Y$ is a homotopy equivalence over Z,*
3. *$f : X \to Y$ is a weak homotopy equivalence over Z,*

Proof Let $f : X \to Y$ be a weak equivalence. Factor $f = tf \circ tc$ with a trivial cofibration $tc : X \rightarrowtail F$ followed by a trivial fibration $tf : F \twoheadrightarrow Y$, both of which are then also over Z.

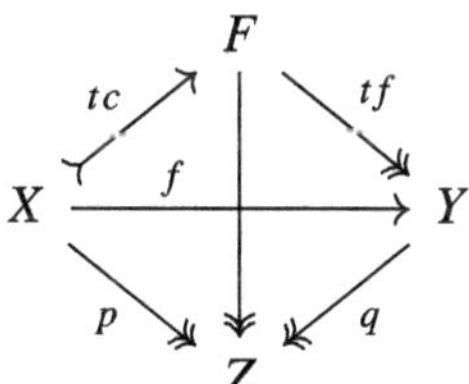

The proof of Proposition 5.19 ($1 \Rightarrow 2$) now applies over Z, *mutatis mutandis*, to show that $tc : X \rightarrowtail F$ is a homotopy equivalence over Z. Similarly, the proof of Lemma 5.9 also works over Z to show that $tf : F \twoheadrightarrow Y$ is a homotopy equivalence over Z. Thus $f = tf \circ tc$ is a homotopy equivalence over Z.

Any homotopy equivalence over Z is clearly a weak homotopy equivalence over Z, by the same proof as for Lemma 5.13 (using the fact that $X \twoheadrightarrow Z$ and $Y \twoheadrightarrow Z$ in order to form the required exponentials).

If $f : X \to Y$ is a weak homotopy equivalence, then factor it as $f = tf \circ c$ with a cofibration $c : X \rightarrowtail F$ followed by a trivial fibration $tf : F \twoheadrightarrow Y$, both of which are also over Z. We thus just need to show that $c : X \rightarrowtail F$ is a trivial cofibration. As in the first step, $tf : F \twoheadrightarrow Y$ is a homotopy equivalence over Z, whence a weak homotopy equivalence by the second step, and so by 3-for-2 for weak homotopy equivalences over Z, Lemma 8.11, $c : X \rightarrowtail F$ is also a weak homotopy equivalence over Z. Now, as in the proof of Proposition 5.19, factor $c = f \circ tc$ as a trivial cofibration $tc : X \rightarrowtail C$ followed by a fibration $f : C \twoheadrightarrow F$, both over Z. By steps 1 and 2, $tc : X \rightarrowtail C$ is then a weak homotopy equivalence over Z. By 3-for-2 for weak homotopy equivalences, Lemma 8.11, $f : C \twoheadrightarrow F$ is also a weak homotopy equivalence over Z. It remains to show that the fibration $f : C \twoheadrightarrow F$ is a weak equivalence. This follows by repeating the reasoning for Lemma 5.18, and the results leading up to it, over Z. □

Using Lemma 8.11 we now have the desired:

Corollary 8.13 *For any object Z, the weak equivalences between fibrations into Z satisfy the 3-for-2 condition.*

Remark 8.14 Our immediate goal has been to show Corollary 8.13, which will be used below to establish the equivalence extension property. The following two results, Lemma 8.15 and Proposition 8.16, will emphatically *not* be used in the sequel, but are included here simply to complete the study of the relative (pre)model structure. They both assume the fibration extension property, Corollary 9.7, which will not be proved until Chap. 9.

Lemma 8.15 *Let $f : X \to Y$ be any map over Z.*

1. *If $f : X \to Y$ is homotopy equivalence over Z, then $Z_! f : Z_! X \to Z_! Y$ is a homotopy equivalence.*
2. *If $X \twoheadrightarrow Z$ and $Y \twoheadrightarrow Z$ are fibrations and $f : X \to Y$ is weak homotopy equivalence over Z, then $Z_! f : Z_! X \to Z_! Y$ is a weak homotopy equivalence.*

Proof (1) is immediate from the fact that $Z_!$ preserves homotopies, Lemma 8.8. For (2), let $f : X \to Y$ be a weak homotopy equivalence over Z between fibrations $X \twoheadrightarrow Z$ and $Y \twoheadrightarrow Z$, and let K be any fibrant object in cSet. Consider the internal precomposition map,

$$K^{Z_! f} : K^{Z_! Y} \to K^{Z_! X} ,$$

which we would like to show is a bijection under $\pi_0 : \mathsf{cSet} \to \mathsf{Set}$. Since $K \twoheadrightarrow 1$ is a fibration, so is its pullback $Z^* K \twoheadrightarrow Z$. Therefore, since $f : X \to Y$ is weak homotopy equivalence over Z, the precomposition map over Z,

$$(Z^* K)^f : (Z^* K)^Y \to (Z^* K)^X ,$$

is bijective on connected components,

$$(\pi_0)_Z\big((Z^* K)^f\big) : (\pi_0)_Z\big((Z^* K)^Y\big) \cong (\pi_0)_Z\big((Z^* K)^X\big) .$$

But now observe that in the coequalizer (8.1.6) that defines $(\pi_0)_Z\big((Z^* K)^X\big)$, we have

$$\mathrm{Hom}_Z(1_Z, (Z^* K)^X) \cong \mathrm{Hom}_Z(X, (Z^* K))$$

$$\cong \mathrm{Hom}(Z_! X, K) \cong \mathrm{Hom}(1, K^{Z_! X}) ,$$

and similarly

$$\mathrm{Hom}_Z(\mathrm{I}_Z, (Z^*K)^X) \cong \mathrm{Hom}_Z(Z^*\mathrm{I} \times X, Z^*K)$$

$$\cong \mathrm{Hom}_Z(\mathrm{I} \times Z_!X, K) \cong \mathrm{Hom}(\mathrm{I}, K^{Z_!X}).$$

Thus $(\pi_0)_Z((Z^*K)^X) \cong (\pi_0)(K^{Z_!X})$, and the same is true with Y in place of X. So $K^{Z_!f} : K^{Z_!Y} \to K^{Z_!X}$ is also bijective on connected components. □

Proposition 8.16 *Let Z be any object in* cSet *and $X \twoheadrightarrow Z$ and $Y \twoheadrightarrow Z$ fibrations. Assuming the fibration extension property, Corollary 9.7, the following conditions are equivalent for any map $f : X \to Y$ over Z.*

1. *$f : X \to Y$ is a weak equivalence over Z.*
2. *$f : X \to Y$ is a homotopy equivalence over Z.*
3. *$f : X \to Y$ is a weak homotopy equivalence over Z.*
4. *$Z_!f : Z_!X \to Z_!Y$ is a weak equivalence.*
5. *$Z_!f : Z_!X \to Z_!Y$ is a homotopy equivalence in* cSet.
6. *$Z_!f : Z_!X \to Z_!Y$ is a weak homotopy equivalence in* cSet.

Proof In Proposition 8.12 we showed the implications $1 \Leftrightarrow 2 \Leftrightarrow 3$. We also have $1 \Leftrightarrow 4$ by definition, and $4 \Rightarrow 6$ and $5 \Rightarrow 6$ by Lemmas 5.16 and 5.13. Moreover, by Lemma 8.15 we have $2 \Rightarrow 5$ and $3 \Rightarrow 6$. Thus all 6 conditions will be equivalent once we have $6 \Rightarrow 4$, which follows from Proposition 5.27 and the fibration extension property, Corollary 9.7. □

8.2 Pathobject Factorizations

For any map $f : X \to Y$ in cSet, recall the *pathobject factorization* $f = t \circ s$ indicated below.

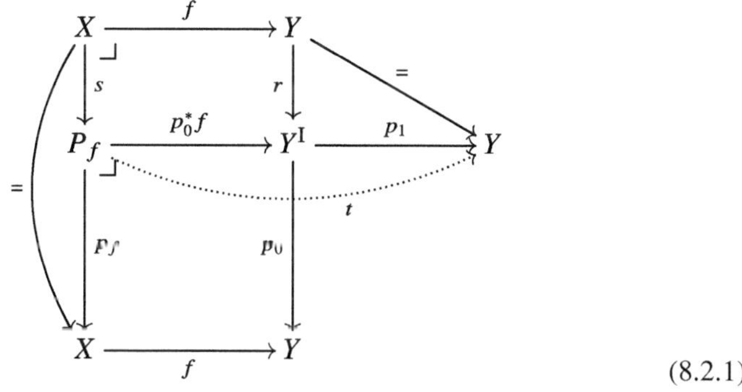

$$(8.2.1)$$

Here p_0, p_1 are the evaluations Y^{δ_0}, $Y^{\delta_1} : Y^\mathrm{I} \to Y$ at the endpoints $\delta_0, \delta_1 : 1 \to \mathrm{I}$, and let $r := Y^!$ for $! : \mathrm{I} \to 1$, so that $p_0 r = p_1 r = 1_Y$. Then let $p_f := f^*p_0 : P_f \to Y$, the pullback of p_0 along f, and $s := f^*r : X \to P_f$ (as a map over X). Finally, let $t := p_1 \circ p_0^* f : P_f \to Y$ be the indicated horizontal composite.

We then have the following facts:

1. The retraction $p_0 \circ r = 1_Y$ pulls back along f to a retraction $p_f \circ s = 1_X$.
2. If Y is a fibrant object, then $p_0, p_1 : Y^{\mathrm{I}} \to Y$ are both trivial fibrations, by Proposition 4.13.
3. If X and Y are both fibrant then $t = p_1 \circ p_0^* f : P_f \to Y$ is a fibration. This can be seen by factoring the maps $p_0, p_1 : Y^{\mathrm{I}} \rightrightarrows Y$ through the product projections as

$$\pi_0 \circ p, \ \pi_1 \circ p : Y^{\mathrm{I}} \to Y \times Y \rightrightarrows Y$$

where $p = (p_0, p_1)$, and then interpolating the pullback $(f, 1_Y) : X \times Y \to Y \times Y$ into (8.2.1) as indicated below.

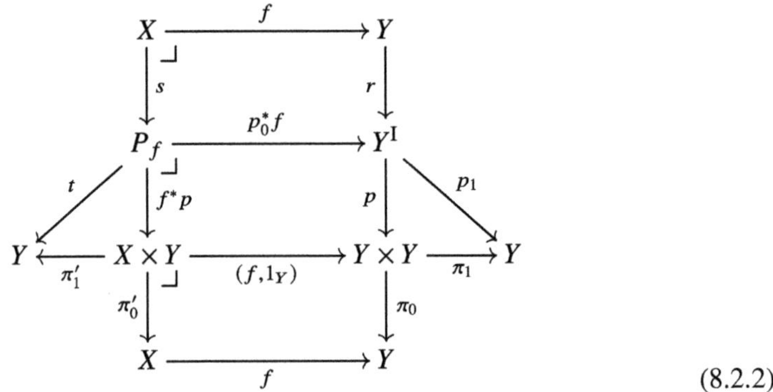

$$(8.2.2)$$

The second factor $t = p_1 \circ p_0^* f : P_f \to Y$ now appears also as $\pi_1 \circ (f, 1_Y) \circ f^* p$, which is equal to the pullback $f^* p : P_f \to X \times Y$ followed by the second projection $\pi_1' : X \times Y \to Y$ (which is not a pullback). But if Y is fibrant, then $p : Y^{\mathrm{I}} \to Y \times Y$ is a fibration by the $\otimes \dashv \Rightarrow$ adjunction, since $p = \partial \Rightarrow Y$ (this is just as in Proposition 4.13, but with the cofibration $\partial : 1 + 1 \rightarrowtail \mathrm{I}$ in place of the trivial cofibration $\delta_\epsilon : 1 \to \mathrm{I}$). Therefore the pullback $f^* p$ is also a fibration. And if X is fibrant, then the second projection $\pi_1' : X \times Y \to Y$ is a fibration. Thus in this case, $t = \pi_1' \circ f^* p : P_f \to Y$ is a fibration, as claimed.

Summarizing (1)–(3):

Lemma 8.17 *For any map $f : X \to Y$ there is a factorization $f = t \circ s$,*

$$(8.2.3)$$

in which:

1. *The map s is a section of a map $p_f : P_f \to X$.*
2. *If Y is fibrant, then p_f is a trivial fibration.*
3. *If both X and Y are fibrant, then t is a fibration.*

Note that the retraction $p_f : P_f \to X$ of s is not over Y.

Next, if $f : X \to Y$ is a map over any base object Z in cSet, we can use the same factorization $f = t \circ s$ to get a factorization in the slice category over Z,

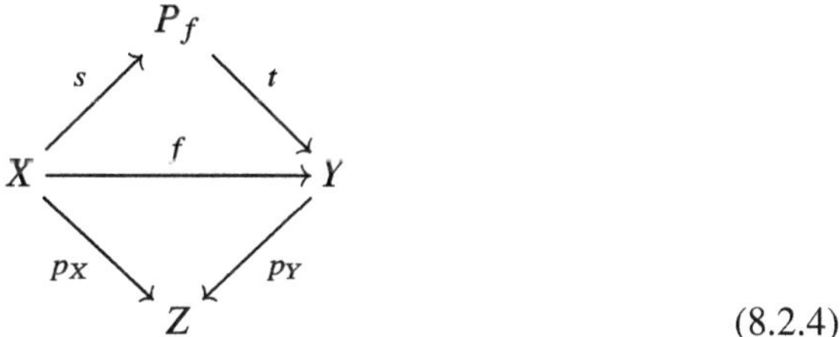

$$(8.2.4)$$

with $p_Y \circ t : P_f \to Z$; however, the maps s, t will no longer have the properties stated in Lemma 8.17, because e.g. $p_0 : Y^I \to Y$ need not be a trivial fibration, even when $p_Y : Y \to Z$ is a fibration, since the object Y need not be fibrant if the base Z is not fibrant.

To remedy this, we can instead build a *fiberwise pathobject factorization* by using the relative pathobject $Y^{I_Z} \to Z$, where the indicated exponential is taken in the slice over Z, and the interval object I_Z occurring in the exponent is the relative one from (8.1.2), i.e. the result of pulling the interval I back from cSet along $Z \to 1$. The pathspace factorization is then constructed as in (8.2.1), but now in the slice cSet$/_Z$, using the pulled back interval $1_Z \rightrightarrows I_Z$. Moreover, the resulting factorization $f = t \circ s : X \to P_f \to Y$ is then stable under pullback along any map $g : Z' \to Z$, in the sense that $g^*(Y^{I_Z}) \simeq g^*(Y)^{I_{Z'}}$ and so $g^* P_f = P_{g^* f}$, where $g^* f : g^* X \to g^* Y$, and similarly for the factors $g^* s$ and $g^* t$.

In more detail, let us review the foregoing steps in the relative case, with reference to the following diagram.

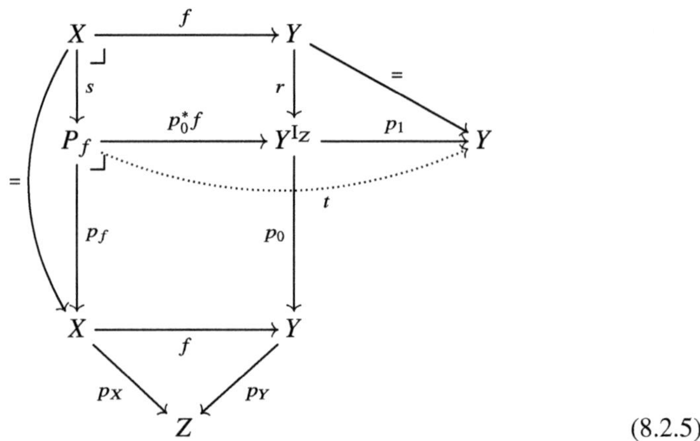

$$(8.2.5)$$

1. The exponential of $p_Y : Y \to Z$, taken in $\mathsf{cSet}/_Z$, by the *constant* maps $Z^*\delta_\epsilon :$ $Z^*1 \to Z^*\mathrm{I}$, which we write as $\delta_\epsilon : 1_Z \to \mathrm{I}_Z$, are now maps $p_\epsilon := Y^{\delta_\epsilon} : Y^{\mathrm{I}_Z} \to Y$ over Z, for $\epsilon = 0, 1$. The retraction $p_0 \circ r = 1_Y$ (with r defined accordingly) is now also over Z, and it still pulls back along f to a retraction $p_f \circ s = 1_X$, also over Z.
2. If $p_Y : Y \twoheadrightarrow Z$ is a fibration, then the maps $p_0, p_1 : Y^{\mathrm{I}_Z} \to Y$ over Z are again trivial fibrations by Lemma 8.6, since these are pullback-homs over Z of the form $\delta_\epsilon \Rightarrow_Z Y$.
3. If $X \twoheadrightarrow Z$ and $Y \twoheadrightarrow Z$ are both fibrations, then for the same reason $t = p_1 \circ p_0^* f :$ $P_f \to Y$ is a fibration.

Again, summarizing (1)–(3) in the relative case:

Lemma 8.18 *For any map* $f : X \to Y$ *over any base* $Z \in \mathsf{cSet}$, *there is a stable factorization* $f = t \circ s$ *over* Z,

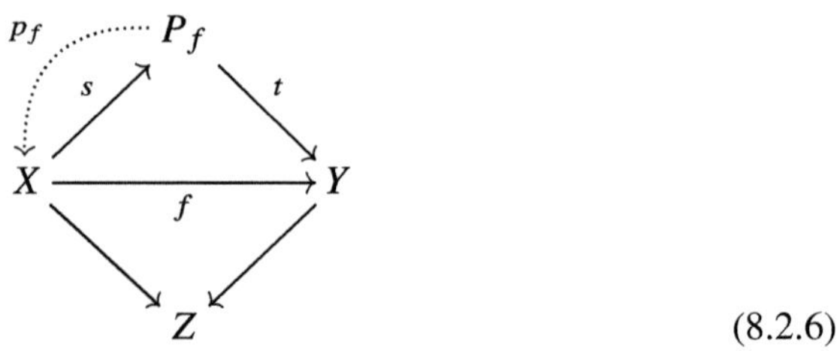

$$(8.2.6)$$

in which:

1. *The map* $s : X \to P_f$ *is a section of a map* $p_f : P_f \to X$ *over* Z.
2. *If* $Y \to Z$ *is a fibration, then* $p_f : P_f \to X$ *is a trivial fibration.*
3. *If both* $X \to Z$ *and* $Y \to Z$ *are fibrations, then* $t : P_f \to Y$ *is a fibration.*

Note that the retraction $p_f : P_f \to X$ *of* s *is not over* Y.

Finally, the following fact concerning just the cofibration weak factorization system will also be needed.

Lemma 8.19 *Let $p : E \twoheadrightarrow B$ be a trivial fibration and $c : C \rightarrowtail B$ a cofibration. Then the unit $\eta : E \to c_* c^* E$ over B of the base change along c,*

$$c^* \dashv c_* : \mathsf{cSet}/C \longrightarrow \mathsf{cSet}/B$$

is also a trivial fibration.

Proof Regarding $c : C \rightarrowtail B$ as a subobject $C \rightarrowtail 1_B$ in cSet/B, the unit map $\eta : E \to c_* c^* E = E^C$ is the pullback-hom $c \Rightarrow_B p$ in the slice category over B, as shown below.

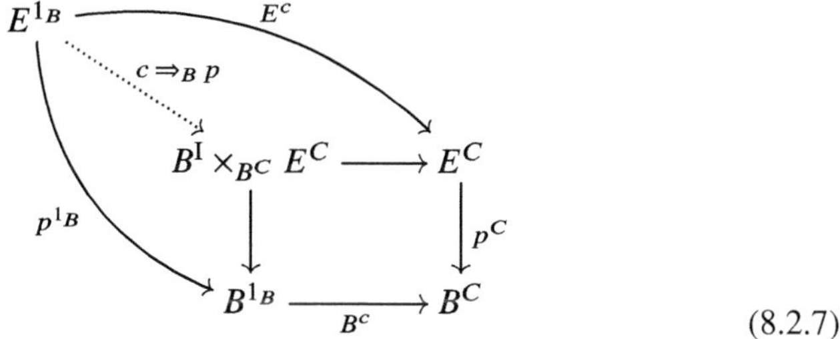

$$(8.2.7)$$

We use the fact that in cSet/B we have $B^c : B^{1_B} \cong 1_B \cong B^C$ and so

$$(c \Rightarrow_B p) = E^c : E \longrightarrow E^C ,$$

which is indeed $\eta : E \to c_* c^* E = E^C$.

Now for any cofibration $A \stackrel{a}{\rightarrowtail} X \to B$ over B, by Lemma 4.2 we have an equivalence of diagonal filling conditions in cSet/B,

$$a \pitchfork (c \Rightarrow_B p) \quad \text{iff} \quad (u \otimes_B c) \pitchfork p.$$

But since $c : C \rightarrowtail B$ is a cofibration, $a \otimes_B c$ is also a cofibration, since $a : A \rightarrowtail X$ is one, and by axiom (C6), cofibrations are closed under pushout-products (also in a slice). Thus $(a \otimes_B c) \pitchfork p$ indeed holds, since p is a trivial fibration. $\square$

Proposition 8.20 (Equivalence Extension Property)   *Weak equivalences extended along cofibrations in the following sense: given a cofibration $c : C' \rightarrowtail C$ and fibrations $A' \twoheadrightarrow C'$ and $B \twoheadrightarrow C$, and a weak equivalence $w' : A' \simeq c^* B$ over C',*

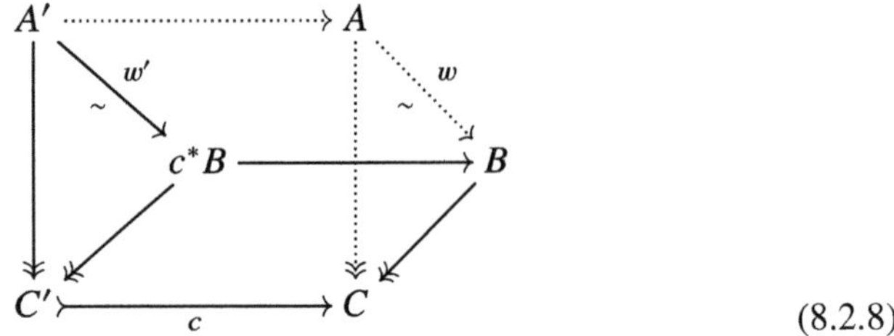

$$(8.2.8)$$

there is a fibration $A \twoheadrightarrow C$ and a weak equivalence $w : A \simeq B$ over C that pulls back along $c : C' \rightarrowtail C$ to w', so $c^* w = w'$.

Proof Call the given fibration $q : B \to C$ and let $b := q^* c : c^* B \to B$ be the indicated pullback, which is thus also a cofibration. Let $w := b_* w' : A \to B$ be the pushforward of w' along b. Composing w with q gives the map $p := q \circ w : A \to C$. Since b is monic, we indeed have $b^* w = w'$, thus filling in all the dotted arrows in (8.2.8). Note moreover that $c^* w = b^* w = w'$, as required. It remains to show that $p : A \to C$ is a fibration and $w : A \to B$ is a weak equivalence.

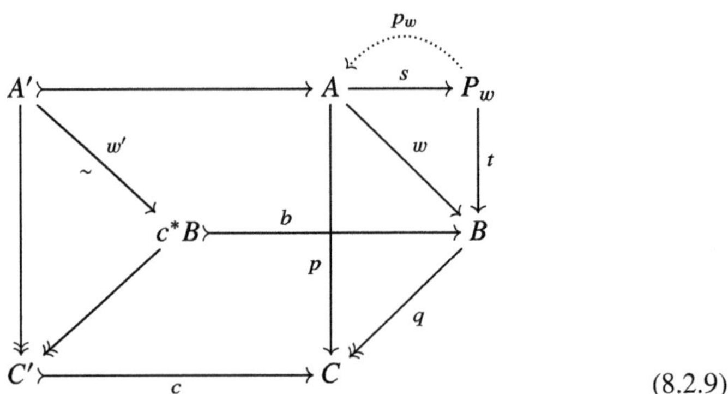

$$(8.2.9)$$

Let us name $p' := c^* p : A' \to C'$ and $B' := c^* B$ and $q' := c^* q$. Now let $w = t \circ s$ be the (relative) pathspace factorization (8.2.5) of w, as a map over C. Since $q : B \to C$ is a fibration, by Lemma 8.18, we know that $s : A \to P_w$ has a retraction $p_w : P_w \to A$ over C which is a trivial fibration.

The pathspace factorization $w = t \circ s : A \to P_w \to B$ is stable under pullback along c, providing a pathspace factorization of $c^* w = w' = t' \circ s' : A' \to P_{w'} \to B'$ over C'. Since both p' and q' are fibrations, the retraction $p_{w'} : P_{w'} \to A'$ is a trivial fibration, and now $t' : P_{w'} \to B'$ is a fibration.

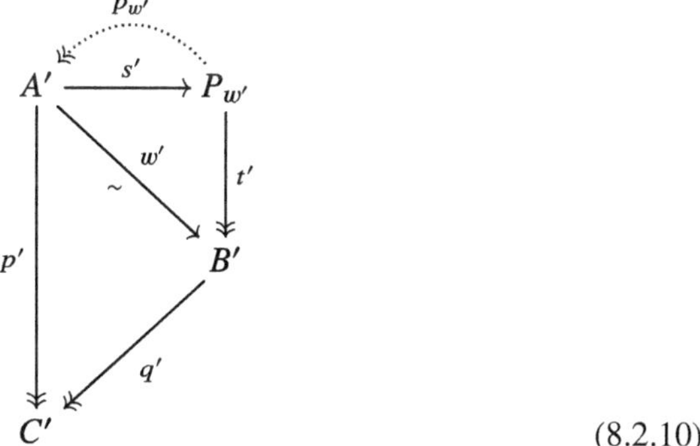

$$(8.2.10)$$

Thus the composite $q' \circ t' : P_{w'} \to B' \to C'$ is a fibration and therefore, by the retraction over C' with the trivial fibration $p_{w'}$, we have that $s' : A' \to P_{w'}$ is a weak equivalence, by 3-for-2 for weak equivalences between fibrations, Corollary 8.13. For the same reason, t' is then a weak equivalence, and therefore a trivial fibration.

Since $t' = c^*t = b^*t$ is a trivial fibration, its pushforward b_*b^*t along b is also one, by Corollary 3.10. Moreover, $b_*b^*t : b_*b^*P_w \to B$ admits a unit $\eta : P_w \to b_*b^*P_w$ (over B).

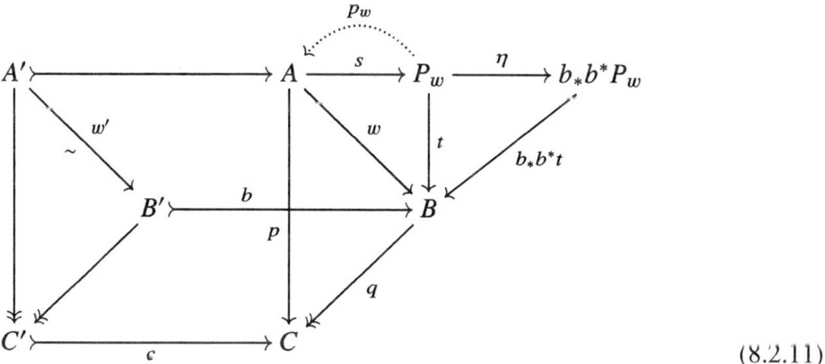

$$(8.2.11)$$

We now *claim* that $\eta : P_w \to b_*b^*P_w$ is a trivial fibration. Given that, the composite $t = b_*b^*t \circ \eta$ is also a trivial fibration, whence $q \circ t : P_w \to C$ is a fibration, and so its retract $p : A \to C$ is a fibration. Moreover, since s is a section of the trivial fibration $p_w : P_w \to A$ between fibrations, again by Corollary 8.13 it is also a weak equivalence. Thus $w = t \circ s$ is a weak equivalence, and we are finished.

To prove the remaining claim that $\eta : P_w \to b_*b^*P_w$ is a trivial fibration, we shall use Lemma 8.19. It does not apply directly, however, since $t : P_w \to B$ is not yet known to be a trivial fibration. Instead, we show that η is a pullback of the corresponding unit at the trivial fibration $p_1 : B^I \to B$.

Consider the following cube (viewed with $b : B' \to B$ at the front).

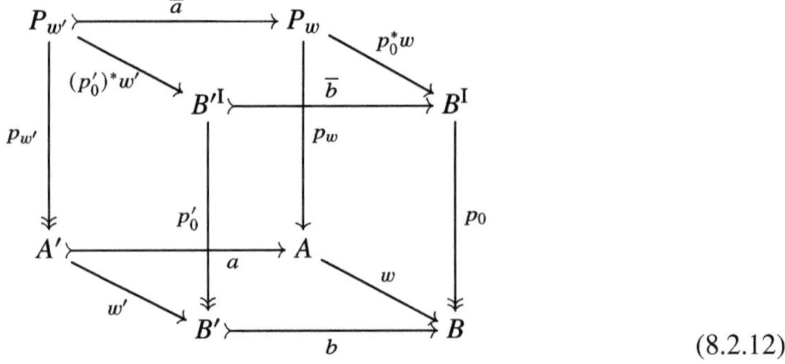

$$(8.2.12)$$

The right hand face is a pullback by definition, and the remainder results from pulling the entire right face back along b, by the stability of the pathspace factorization, Lemma 8.18. Thus all faces in the cube (8.2.12) are pullbacks. The base is also a pushforward, $b_* w' = w$, again by definition. Thus the top face is also a pushforward, $\overline{b}_*((p_0')^* w') = p_0^* w$. Indeed, since the front face is a pullback, the Beck-Chevalley condition applies, and so we have $\overline{b}_*(p_0')^*(w') = p_0^* b_*(w') = p_0^* w$.

Now consider the following, in which the top square remains the same as in (8.2.12), but p_0 has been replaced by $p_1 : B^I \to B$, so the composite at right is by definition $t = p_1 \circ p_0^* w$.

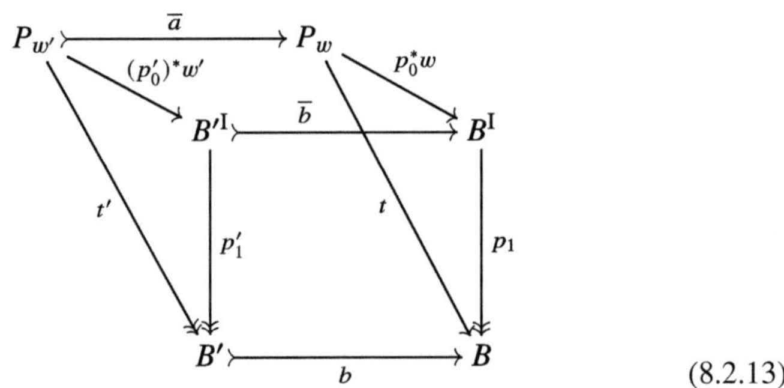

$$(8.2.13)$$

The horizontal direction is still pullback along b; let us rename $p_0^* w =: u$ so that $(p_0')^* w' = b^* u$ and $t' = b^* t$ and $p_1' = b^* p_1$ to make this clear. We then add the pushforward along b on the right, in order to obtain the two units η.

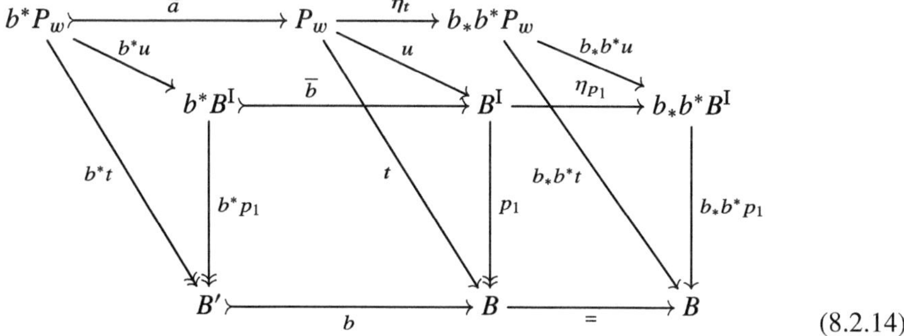

$$(8.2.14)$$

By the usual calculation of pushforwards in slice categories, we have $\overline{b}_* \cong \eta_{p_1}^* \circ b_*$, and so for b^*u we have $\overline{b}_* b^*u = \eta_{p_1}^* b_* b^*u$. But as we just determined in (8.2.12) the top left square is already a pushforward, and therefore $u = \eta_{p_1}^* b_* b^*u$, so the top right naturality square is a pullback.

To finish the proof as planned, $p_1 : B^{\mathrm{I}} \to B$ is a trivial fibration because $q : B \to C$ is a fibration, and $b : B' \rightarrowtail B$ is a cofibration because it is a pullback of $c : C' \rightarrowtail C$. Thus by Lemma 8.19, we have that $\eta_{p_1} : B^{\mathrm{I}} \to b_* b^* B^{\mathrm{I}}$ is a trivial fibration, and so its pullback $\eta_t : P_w \to b_* b^* P_w$ is a trivial fibration, as claimed. $\square$

Remark 8.21 Note that $p : A \to C$ is small if $q : B \to C$ is small.

Chapter 9
The Fibration Extension Property

Given a universal fibration $\dot{\mathcal{U}} \twoheadrightarrow \mathcal{U}$, such as $\mathsf{\dot{Fib}} \twoheadrightarrow \mathsf{Fib}$ of Proposition 7.17, the fibration extension property (Definition 5.24) is closely related to the statement that the base object $\mathcal{U}$ is fibrant. For Kan simplicial sets, Voevodsky proved the latter directly, using the theory of minimal fibrations, cf. [50]. In a more general (but still simplicial) setting, [67] gives a proof using univalence, in the form of the equivalence extension property of Chap. 8, but that proof also uses the 3-for-2 property for weak equivalences, which we do not yet have. For cubical sets, [28] uses the equivalence extension property to prove that $\mathcal{U}$ is fibrant without assuming 3-for-2 for weak equivalences, via a neat type theoretic argument reducing box filling to an operation of *Kan-composition*. We shall prove that $\mathcal{U}$ is fibrant in the category cSet using the equivalence extension property, also without assuming 3-for-2 for weak equivalences, but via a different argument than that in [28] not using (type theory or) Kan composition.

9.1 Fibrancy of the Universe

Returning to the relation between the fibration extension property and the fibrancy of the base object of the universal fibration $\dot{\mathcal{U}} \twoheadrightarrow \mathcal{U}$, it is easy to see that the latter implies the former. Indeed, let $t : X \rightarrowtail X'$ be a trivial cofibration and $Y \twoheadrightarrow X$ a (small) fibration. To extend Y along t, take a classifying map $y : X \to \mathcal{U}$, so that $Y \cong y^* \dot{\mathcal{U}}$ over X. If $\mathcal{U}$ is fibrant then we can extend y along $t : X \rightarrowtail X'$ to get $y' : X' \to \mathcal{U}$ with $y = y' \circ t$. The pullback $Y' = (y')^* \dot{\mathcal{U}} \twoheadrightarrow X'$ is then a (small) fibration such that $t^* Y' \cong t^*(y')^* \dot{\mathcal{U}} \cong y^* \dot{\mathcal{U}} \cong Y$ over X.

© The Author(s), under exclusive license to Springer Nature Switzerland AG 2026

S. Awodey, *Cartesian Cubical Model Categories*, Lecture Notes in Mathematics 2385, https://doi.org/10.1007/978-3-032-08730-0_9

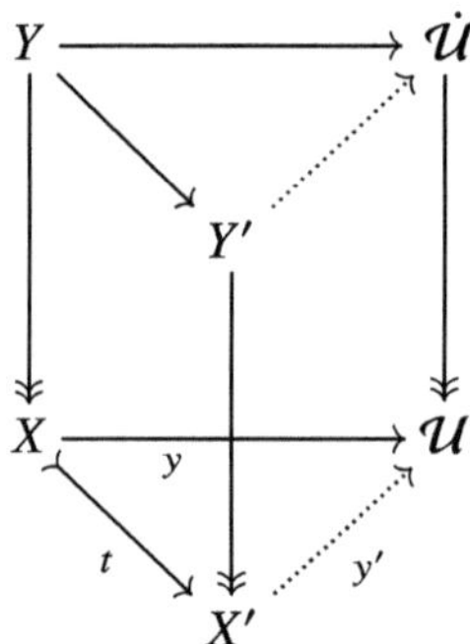

In this way, we have:

Proposition 9.1 *If the base object $\mathcal{U}$ of the universal fibration $\dot{\mathcal{U}}\twoheadrightarrow\mathcal{U}$ in* cSet *is fibrant, then the fibration weak factorization system has the fibration extension property.*

Proof In more detail, we assume that in the cofinal sequence of universes $\mathcal{U}_\alpha \subset \mathcal{U}_\beta \ldots$ coming from Remark 7.9, all of the objects $\mathcal{U}_\kappa$ are fibrant. Let $t : X\rightarrowtail X'$ be a trivial cofibration and $Y\twoheadrightarrow X$ a fibration. By our standing assumption regarding size, $Y\twoheadrightarrow X$ is κ-small for some κ. There is therefore a classifying map $y : X\rightarrow\mathcal{U}_\kappa$ with $Y \cong y^*\dot{\mathcal{U}}_\kappa$ over X. Since $\mathcal{U}_\kappa$ is fibrant, we can then extend y along $t : X\rightarrowtail X'$ to get $y' : X'\rightarrow\mathcal{U}_\kappa$ with $y = y' \circ t$. The pullback $Y' = (y')^*\dot{\mathcal{U}}_\kappa\twoheadrightarrow X'$ is then a (κ-small) fibration such that $t^*Y' \cong t^*(y')^*\dot{\mathcal{U}}_\kappa \cong y^*\dot{\mathcal{U}}_\kappa \cong Y$ over X, as required. $\square$

Remark 9.2 An alternative approach would be to give a direct proof of the fibration extension property of Definition 5.24, without going via the universe. This is done in a different setting in [66].

Conversely, given the Realignment Lemma 7.22, the fibration extension property also implies the fibrancy of $\mathcal{U}$:

Corollary 9.3 *The fibration extension property implies that the base $\mathcal{U}$ of the universal fibration $\dot{\mathcal{U}}\twoheadrightarrow\mathcal{U}$ is fibrant: given any $y : X\rightarrow\mathcal{U}$ and trivial cofibration $t : X\rightarrowtail X'$, there is a map $y' : X'\rightarrow\mathcal{U}$ with $y' \circ t = y$.*

Proof Take the pullback of $\dot{\mathcal{U}}\twoheadrightarrow\mathcal{U}$ along $y : X\rightarrow\mathcal{U}$ to get a (small) a fibration $Y\twoheadrightarrow X$, which extends along the (trivial) cofibration $t : X\rightarrowtail X'$ by the fibration extension property, to a (small) fibration $Y'\twoheadrightarrow X'$ with $Y \cong t^*Y'$ over X. By realignment there is a classifying map $y' : X'\rightarrow\mathcal{U}$ for Y' with $y' \circ t = y$. $\square$

Now let us show the following.

Proposition 9.4 *The base $\mathcal{U}$ of the universal fibration $\dot{\mathcal{U}}\twoheadrightarrow\mathcal{U}$ in* cSet, *as constructed in Sect. 7.3, is a fibrant object.*

Proof By Corollary 4.8, $\mathcal{U}$ is an unbiased fibrant object if the canonical map $u = \langle \mathsf{p}_2, \mathsf{eval}\rangle$ in the following diagram in cSet, is a trivial fibration.

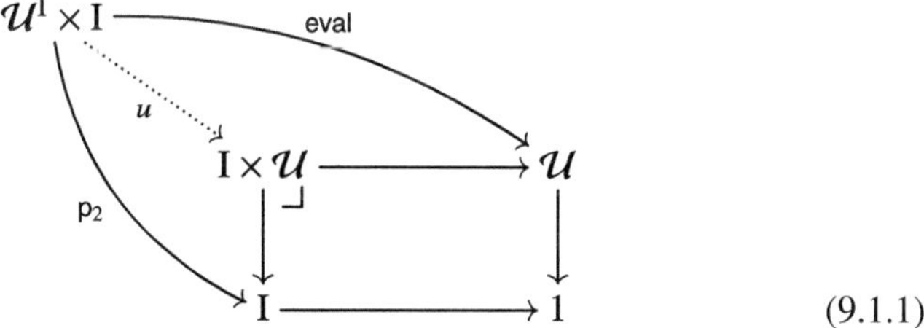

$$\tag{9.1.1}$$

Thus consider a filling problem of the following form, with an arbitrary cofibration $c : C \rightarrowtail Z$.

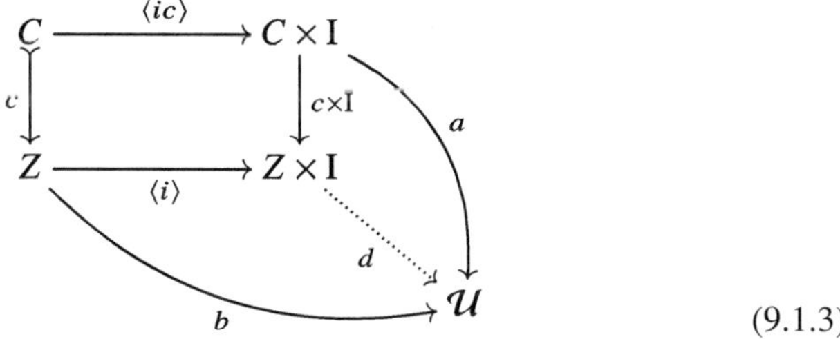

$$\tag{9.1.2}$$

The horizontal maps may be written in the form $\langle i, b \rangle : Z \to I \times \mathcal{U}$ and $\langle \tilde{a}, ic \rangle : C \to \mathcal{U}^I \times I$, regarding $i : Z \to I$ as an I-indexing.

Transposing $\tilde{a}$ to $a : C \times I \to \mathcal{U}$ we obtain the new problem

$$\tag{9.1.3}$$

in which we recall from (4.4.4) the notation $\langle i \rangle = \langle 1_Z, i \rangle : Z \to Z \times I$ for the graph of a map $i : Z \to I$. Given a map d as shown in (9.1.3), we can obtain the indicated diagonal filler in (9.1.2) as $\langle \tilde{d}, i \rangle : Z \to \mathcal{U}^I \times I$.

As a sanity check, note that $b \circ c = a \circ \langle ic \rangle$ turns the problem (9.1.3) into that of extending the copair $[b, a]$ along the unique map

$$Z +_C (C \times I) \longrightarrow Z \times I,$$

which is exactly the (trivial cofibration) pushout-product $c \otimes_i \delta$ from (4.4.5), recalled below for the reader's convenience.

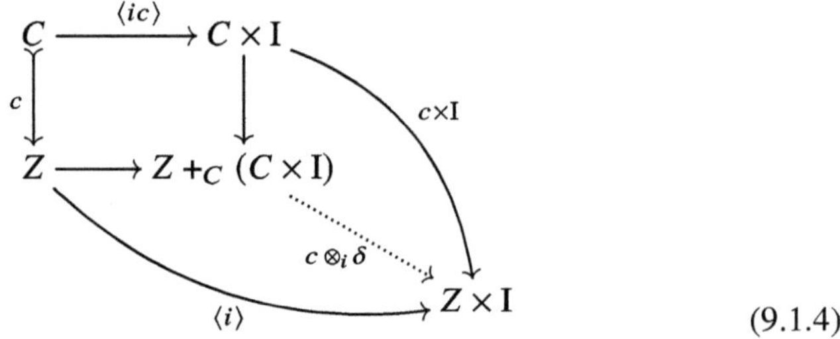

$$(9.1.4)$$

Returning to (9.1.3), take pullbacks of $\dot{\mathcal{U}} \twoheadrightarrow \mathcal{U}$ along a and b to get fibrations $p_a : A \twoheadrightarrow C \times I$ and $p_b : B \twoheadrightarrow Z$ respectively, and let

$$p_c := \langle ic \rangle^* p_a : A_c \longrightarrow C$$

be the corresponding "fiber of A over the graph of ic". We then have $c^* B \cong A_c$ over C by the commutativity of the outer square of (9.1.2).

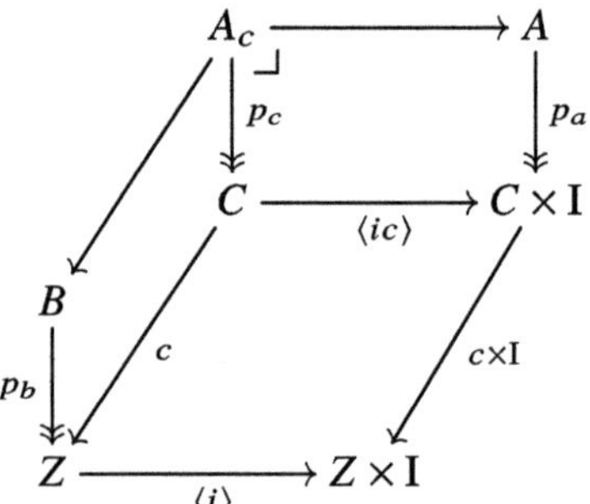

The diagonal filler sought in (9.1.2) now corresponds, again by transposition and pullback of $\dot{\mathcal{U}} \twoheadrightarrow \mathcal{U}$, to a fibration $p_d : D \twoheadrightarrow Z \times I$ with $\langle i \rangle^* D \cong B$ over Z and $(c \times I)^* D \cong A$ over $C \times I$, as indicated below.

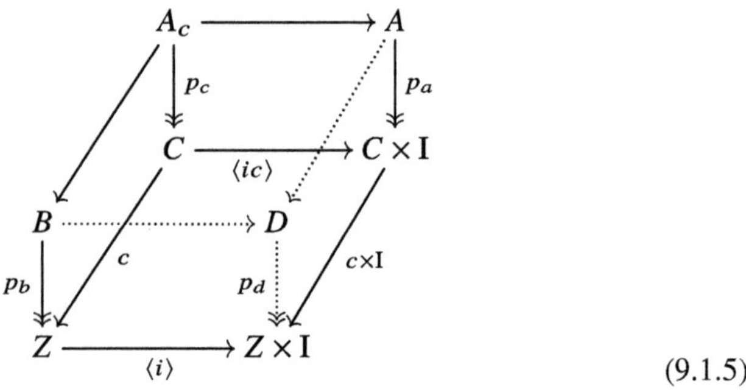

$$(9.1.5)$$

We shall construct $p_d : D \twoheadrightarrow Z \times I$ using the equivalence extension property (Proposition 8.20) as follows. First apply the functor $(-) \times I$ to the left vertical (pullback) face of the cube in (9.1.5) to get the following, with a new pullback square on the right with the indicated fibrations.

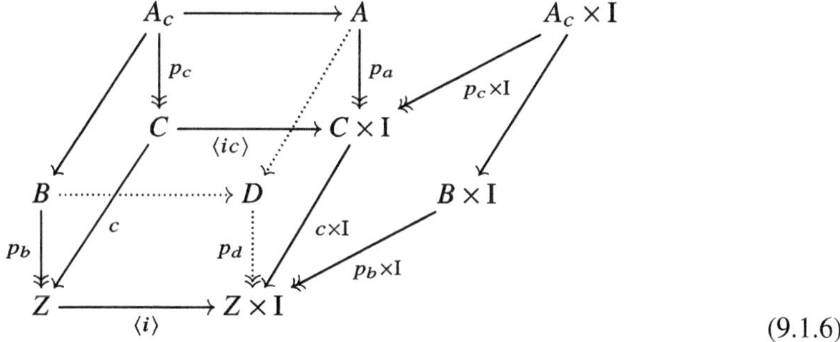

$$(9.1.6)$$

We now *claim* that there is a weak equivalence $e : A \simeq A_c \times I$ over $C \times I$. From this it follows by the equivalence extension property (Proposition 8.20) that there are:

(i) a fibration $p_d : D \twoheadrightarrow Z \times I$ with $(c \times I)^* D \cong A$ over $C \times I$, and
(ii) a weak equivalence $f : D \simeq B \times I$ over $Z \times I$ with $(c \times I)^* f \cong e$ over $C \times I$.

It then remains only to show that $B \cong \langle i \rangle^* D$ over Z to complete the proof.
    To obtain the claimed weak equivalence e, consider the following square,

$$
\begin{array}{ccc}
A_c & \xrightarrow{\langle icp_c \rangle} & A_c \times I \\
\downarrow & & \downarrow{\scriptstyle p_c \times I} \\
A & \xrightarrow[p_a]{} & C \times I,
\end{array}
$$

$$(9.1.7)$$

in which the top horizontal map is the graph of the composite,

$$
A_c \xrightarrow{p_c} C \xrightarrow{c} Z \xrightarrow{i} I,
$$

and the others are the evident ones from (9.1.6). The square is easily seen to commute, and the top map is a trivial cofibration (by Remark 4.12), because it is the graph of a map into I. The left map is also a trivial cofibration by Frobenius (Proposition 6.5), because by its definition in (9.1.5) it is the pullback of another such graph $\langle ic \rangle$ along the fibration p_a. A simple lemma (Lemma 9.5 below) provides the claimed weak equivalence $e : A \simeq A_c \times I$ over $C \times I$.
    To see that $B \cong \langle i \rangle^* D$ over Z, recall from the proof of the equivalence extension property that the map $f : D \simeq B \times I$ is the pushforward of $e : A \simeq A_c \times I$ along the cofibration $b_c \times I : A_c \times I \rightarrowtail B \times I$, where we are calling the evident map in (9.1.6) $b_c : A_c \rightarrowtail B$. Thus by construction $f = (b_c \times I)_* e$. We can then apply the Beck-Chevalley condition for the pushforward using the pullback square on the left below.

$$\begin{array}{ccc}
A_c \rightarrowtail \!\!\xrightarrow{\langle icp_c\rangle}\!\! & A_c \times \mathrm{I} \xleftarrow{\ e\ } A \\
\scriptstyle b_c \downarrow \qquad \lrcorner & \qquad \downarrow \scriptstyle b_c\times\mathrm{I} \\
B \rightarrowtail \!\!\xrightarrow[\langle ip_b\rangle]{}\!\! & B \times \mathrm{I} \xleftarrow{\ f\ } D
\end{array} \qquad (9.1.8)$$

The pullback of e along the top of the square is the identity on A_c, as can be seen by pulling back e as a map over $C \times \mathrm{I}$ along $\langle ic\rangle : C \to C \times \mathrm{I}$. Thus the same is true up to isomorphism for the pullback of f along the bottom.

An application of the Realignment Lemma 7.22 along the trivial cofibration $c \otimes_i \delta$ completes the proof. $\square$

Lemma 9.5 *Suppose the following square commutes and the indicated cofibrations are trivial.*

$$\begin{array}{ccc}
A & \rightarrowtail & C \\
\downarrow & & \downarrow \\
B & \twoheadrightarrow & D
\end{array} \qquad (9.1.9)$$

Then there is a weak equivalence $e : B \simeq C$ over D (and under A).

Proof Use the fact that any two diagonal fillers are homotopic to get a homotopy equivalence $e : B \simeq C$ filling the square. $\square$

Remark 9.6 The foregoing proof of Proposition 9.4, the fibrancy of the universe $\mathcal{U}$, also works, *mutatis mutandis*, for the universe of *biased* fibrations, as used in the setting of [28].

Applying Proposition 9.1 now yields the following.

Corollary 9.7 (Fibration Extension Property)   *The fibration weak factorization system has the fibration extension property (Definition 5.24).*

By Theorem 5.28, finally, we have the following.

Theorem 9.8 *Let C be a class of maps in the category* cSet *of cubical sets satisfying the axioms (C0)-(C8). There is a Quillen model structure $(C, \mathcal{W}, \mathcal{F})$ on* cSet *with:*

1. *The cofibrations are the maps in C.*
2. *The fibrations $\mathcal{F}$ are the maps $f : Y \to X$ for which the canonical map*

$$(f^{\mathrm{I}} \times \mathrm{I}, \mathsf{eval}_Y) : Y^{\mathrm{I}} \times \mathrm{I} \longrightarrow (X^{\mathrm{I}} \times \mathrm{I}) \times_X Y$$

   lifts on the right against C.
3. *The weak equivalences $\mathcal{W}$ are the maps $w : X \to Y$ for which the internal precomposition $K^w : K^Y \to K^X$ is bijective on connected components for every fibrant object K.*

Axioms (C0)–(C8) for cofibrations are satisfied, e.g., by the class of *all* monomorphisms, in which case there are at least two different model structures on cSet, namely, the one just stated in Theorem 9.8, and the test model structure, which is known to be different (see [13]). The axioms (C0)–(C8) are collected (and slightly simplified and renumbered) in Appendix A, where a non-trivial example is also provided. Moreover, we note that the theorem holds equally for any other cube category with finite products, not only the initial one, such as the Dedekind cubes.

Remark 9.9 Observe that, in terms of the universal fibration $\dot{\mathcal{U}} \twoheadrightarrow \mathcal{U}$ constructed in Chap. 7, the equivalence extension property Proposition 8.20 says that the second projection from the classifying type of equivalences $A \simeq B$ between small families,

$$\pi_2 : \Sigma_{A,B} \mathsf{Eq}(A, B) \longrightarrow \mathcal{U},$$

is a trivial fibration. From this, it follows that the canonical transport map

$$* : \mathcal{U}^{\mathrm{I}} \longrightarrow \Sigma_{A,B} \mathsf{Eq}(A, B) \tag{9.1.10}$$

is an equivalence over the base $\mathcal{U}$ via $p_2 : \mathcal{U}^{\mathrm{I}} \to \mathcal{U}$, which is also a trivial fibration because $\mathcal{U}$ is fibrant by Proposition 9.4. In type theory, the pathobject $\mathcal{U}^{\mathrm{I}}$ of course interprets the identity type $A = B$, so the equivalence (9.1.10) can be expressed in the form

$$(A = B) \simeq (A \simeq B).$$

Appendix A
Axioms for Cartesian Cofibrations

A system of maps satisfying the axioms (C0)–(C8) of the main text for the cofibrations in a cartesian cubical model category may be called *Cartesian cofibrations*. Eliminating some redundancy, the axioms can be restated equivalently as follows.

(A0) All cofibrations are monomorphisms.
(A1) All isomorphisms are cofibrations.
(A2) The composite of two cofibrations is a cofibration.
(A3) Any pullback of a cofibration is a cofibration.
(A4) The join of two cofibrant subobjects is a cofibration.
(A5) The diagonal of the interval $I \to I \times I$ is a cofibration.
(A6) Cofibrations are preserved by the pathobject functor $(-)^I$.
(A7) The category of cofibrations and cartesian squares has a terminal object.

Example A.1 Consider the Cartesian cubical presheaves $c\mathcal{E} = \mathcal{E}^{\square^{\mathrm{op}}} = \mathcal{E}^{\mathbb{B}}$ in a topos $\mathcal{E}$. For such (internal) discrete opfibrations (A,α),

$$
\begin{array}{ccccc}
A & \xleftarrow{\ \alpha\ } & \mathbb{B}_1 \times_{\mathbb{B}_0} A & \longrightarrow & A \\
\downarrow & & \downarrow \quad \lrcorner & & \downarrow \\
\mathbb{B}_0 & \xleftarrow[\mathrm{cod}]{} & \mathbb{B}_1 & \xrightarrow[\mathrm{dom}]{} & \mathbb{B}_0
\end{array}
$$

over the (internal) category $\mathbb{B} = (\mathbb{B}_1 \rightrightarrows \mathbb{B}_0)$ of finite bipointed sets, call a subpresheaf $c : (C,\gamma) \to (A,\alpha)$ *locally complemented* if the underlying map $c : C \to A$ over $\mathbb{B}_0 = \mathbb{N}$ is a complemented subobject in $\mathcal{E}_{/\mathbb{N}}$, i.e. $C + \neg C \cong A$ over $\mathbb{N}$. Internally, this means that

$$
C_n + \neg(C_n) \cong A_n \qquad \text{for all } n \in \mathbb{N}, \tag{A.1}
$$

which is a weaker condition than $(A,\alpha) + \neg(A,\alpha) \cong (B,\beta)$ as presheaves (unless $\mathcal{E} = \mathsf{Set}$, in which case it is trivial).

© The Author(s), under exclusive license to Springer Nature Switzerland AG 2026
S. Awodey, *Cartesian Cubical Model Categories*, Lecture Notes
in Mathematics 2385, https://doi.org/10.1007/978-3-032-08730-0

Proposition A.2 *For any topos $\mathcal{E}$, the locally complemented subobjects in the category $\mathsf{c}\mathcal{E}$ of cubical $\mathcal{E}$-objects satisfy the axioms for Cartesian cofibrations.*

Proof Axioms (A0)–(A4) are satisfied by the complemented subobjects in $\mathcal{E}/_\mathbb{N}$, and the forgetful functor $U : \mathsf{c}\mathcal{E} \to \mathcal{E}/_\mathbb{N}$ creates the monos, isos, composites, pullbacks, and joins in question. For (A5), we use the fact that the equality relation on $\mathbb{N}$ is decidable to infer that, for each $[n] = \{0, x_1, \ldots, x_n, 1\}$, the finite set $\{i = j \,|\, 0 \leq i, j \leq n+1\}$ is complemented in $\mathbb{B}([1], [n]) \times \mathbb{B}([1], [n])$, and so for the subpresheaf $\delta : \mathrm{I} \to \mathrm{I} \times \mathrm{I}$ we indeed have,

$$\delta_n + \neg(\delta_n) \cong \mathrm{I}_n \times \mathrm{I}_n \qquad \text{for all } n \in \mathbb{N}.$$

For (A6) we use the fact that the pathobject A^{I} is a shift by one dimension, $(A^{\mathrm{I}})_n = A_{n+1}$, together with (A.1). The cofibration classifier in (A7) is given by applying the right adjoint $U \dashv R : \mathcal{E}/_\mathbb{N} \to \mathsf{c}\mathcal{E}$ to the complemented subobject classifier $\mathbb{N}{*}2$ of $\mathcal{E}/_\mathbb{N}$. $\square$

Appendix B
Cartesian Cubical Sets Classifies Intervals

Recall from Chap. 2 that the objects of the *Cartesian cube category* $\square$ may be taken concretely to be finite, strictly bipointed sets, written

$$[n] = \{0, x_1, \ldots, x_n, 1\},$$

and the arrows $f : [n] \to [m]$ to be all bipointed maps $[m] \to [n]$ (note the direction). The category of *Cartesian cubical sets* is then the presheaf topos

$$\mathsf{cSet} = \mathsf{Set}^{\square^{\mathrm{op}}}.$$

It is generated by the geometric *n-cubes* $\mathrm{I}^n = \mathsf{y}[n]$, with $1 = \mathrm{I}^0$, $\mathrm{I} = \mathsf{y}[1]$; and $\mathrm{I}^n \times \mathrm{I}^m \cong \mathrm{I}^{n+m}$, by preservation of products by the Yoneda embedding $\mathsf{y} : \square^{\mathrm{op}} \hookrightarrow \mathsf{cSet}$. For a cubical set $X : \square^{\mathrm{op}} \to \mathsf{Set}$ we have the usual Yoneda correspondence for the set X_n of *n-cubes in X*,

$$\{c \in X_n\} \cong \{c : \mathrm{I}^n \to X\}.$$

In particular, $\mathrm{I}^n_m = \square([m], [n])$ is the set of *m-cubes in the n-cube*.[1]

---

[1] Note that the cardinality of I^n_m is therefore just $(m + 2)^n$, in comparison to the *Dedekind* cubes $\square_{\wedge,\vee}$ used in [28, 60], for which e.g. the Hom-set $\square_{\wedge,\vee}([n], [1])$ is the *nth Dedekind number*, the number of elements in the free distributive lattice on n generators, which is in general a number so large that it is unknown for values of $n > 8$.

© The Author(s), under exclusive license to Springer Nature Switzerland AG 2026 123
S. Awodey, *Cartesian Cubical Model Categories*, Lecture Notes
in Mathematics 2385, https://doi.org/10.1007/978-3-032-08730-0

Proposition B.1 *The category* cSet *of Cartesian cubical sets is the classifying topos for* intervals*: objects I with points $i, j : 1 \rightrightarrows I$ the pullback of which is* 0*:*

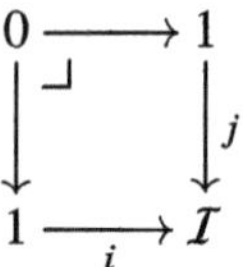

Proof Consider the covariant presentation $\mathsf{cSet} = \mathsf{Set}^{\mathbb{B}}$ where $\mathbb{B}$ is the category of finite, strictly bipointed sets and bipointed maps. We can extend $\mathbb{B} \hookrightarrow \mathbb{B}_=$ by freely adjoining coequalizers, making $\mathbb{B}_=$ the free finite *colimit* category on a co-bipointed object. A concrete presentation of $\mathbb{B}_=$ is the finite bipointed sets, including those with $0 = 1$. Let us write (n) for the bipointed set $\{x_1, \ldots, x_n, *\}$, with n (non-constant) elements and a further element $0 = * = 1$. There is an evident coequalizer $[1] \rightrightarrows [n] \to (n)$, which (only) identifies the distinguished points, and every coqualizer in $\mathbb{B}_=$ has either the form $[m] \rightrightarrows [n] \to [k]$ or $[m] \rightrightarrows [n] \to (k)$, for a suitable choice of k. Note that there are no maps of the form $(m) \to [n]$, and that every map $[m] \to (n)$ factors uniquely as $[m] \to (m) \to (n)$ with $[m] \to (m)$ the canonical coequalizer of 0 and 1. The category $\mathbb{B}_=$ can therefore be decomposed into two "levels", the upper one of which is essentially $\mathbb{B}$, and the lower one consisting of just the objects (n), and thus essentially the finite *pointed* sets, and for each n, there is the canonical coequalizer $[n] \to (n)$ going from the upper level to the lower one.

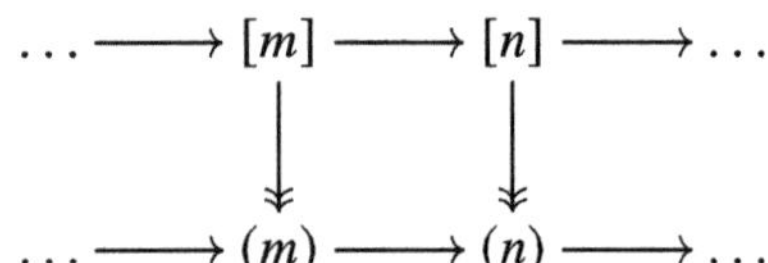

Write $u : \mathbb{B} \to \mathbb{B}_=$ for the upper inclusion, which is the classifying functor of generic co-bipointed object in $\mathbb{B}_=$.

Now consider the induced geometric morphism:

$$\mathsf{Set}^{\mathbb{B}} \; \underset{u_!}{\overset{u_*}{\underset{\longleftarrow u^* \longrightarrow}{\rightrightarrows}}} \; \mathsf{Set}^{\mathbb{B}_=} \qquad u_! \dashv u^* \dashv u_*$$

Since u^* is the restriction along u, the right adjoint u_* must be "prolongation by 1",

$$u_*(P)[n] = P[n],$$

$$u_*(P)(n) = \{*\},$$

with the obvious maps,

$$\cdots \longrightarrow P[m] \longrightarrow P[n] \longrightarrow \cdots$$
$$\downarrow \qquad\qquad \downarrow$$
$$\cdots \longrightarrow \{*\} \longrightarrow \{*\} \longrightarrow \cdots$$

as is easily seen by considering maps in $\mathsf{Set}^{\mathbb{B}_=}$ of the form

$$Q[n] \longrightarrow P[n]$$
$$\downarrow \qquad\qquad \downarrow$$
$$Q(n) \longrightarrow \{*\}.$$

Since $u_* : \mathsf{Set}^{\mathbb{B}} \to \mathsf{Set}^{\mathbb{B}_=}$ is evidently full and faithful, it is the inclusion part of a sheaf subtopos $\mathsf{sh}(\mathbb{B}_=^{\mathrm{op}}, j) \hookrightarrow \mathsf{Set}^{\mathbb{B}_=}$ for a suitable Grothendieck topology j on $\mathbb{B}_=^{\mathrm{op}}$. We claim that j is the closed complement topology of the subobject $[0 = 1] \rightarrowtail 1$ represented by the coequalizer $[0] \to (0)$. Indeed, in $\mathsf{Set}^{\mathbb{B}_=}$ we have the representable functors:

$$\mathrm{I} = \mathsf{y}[1],$$
$$1 = \mathsf{y}[0],$$
$$[0 = 1] = \mathsf{y}(0)$$

fitting into an equalizer $[0 = 1] \to 1 \rightrightarrows \mathrm{I}$, which is the image under Yoneda of the canonical coequalizer $[1] \rightrightarrows [0] \to (0)$ in $\mathbb{B}_=$. The closed complement topology for $[0 = 1] \rightarrowtail 1$ is generated by the single cover $0 \to [0 = 1]$, which can be described logically as forcing the sequent $(0 = 1 \vdash \bot)$ to hold. Recall from [44], Proposition 3.53, the following simple characterization of the sheaves for the closed complement topology of an object $U \rightarrowtail 1$: an object X is a sheaf iff $X \times U \cong U$. In the present case, it therefore suffices to show that for any $P : \mathbb{B}_= \to \mathsf{Set}$ we have:

$$P \times [0 = 1] \cong [0 = 1] \quad \text{iff} \quad P(n) = 1 \text{ for all } n.$$

For any object $b \in \mathbb{B}_=$, consider the map

$$\mathrm{Hom}(\mathsf{y}b, P \times [0 = 1]) \cong \mathrm{Hom}(\mathsf{y}b, P) \times \mathrm{Hom}(\mathsf{y}b, [0 = 1]) \to \mathrm{Hom}(\mathsf{y}b, [0 = 1]).$$

If $b = [k]$, then $\mathrm{Hom}(\mathbf{y}b, [0 = 1]) \cong \mathrm{Hom}_{\mathbb{B}_=}((0), [k]) \cong 0$, and so we always have an iso

$$\mathrm{Hom}(\mathbf{y}b, P \times [0 = 1]) \cong \mathrm{Hom}(\mathbf{y}b, P) \times \mathrm{Hom}(\mathbf{y}b, [0 = 1])$$
$$\cong \mathrm{Hom}(\mathbf{y}b, P) \times 0 \cong 0.$$

If $b = (k)$, then $\mathrm{Hom}(\mathbf{y}b, [0 = 1]) \cong \mathrm{Hom}_{\mathbb{B}_=}((0), (k)) \cong 1$, and we have an iso

$$\mathrm{Hom}(\mathbf{y}b, P \times [\bot = \top]) \cong \mathrm{Hom}(\mathbf{y}(k), P) \times \mathrm{Hom}(\mathbf{y}(k), [0 = 1])$$
$$\cong \mathrm{Hom}(\mathbf{y}(k), P) \times 1 \cong \mathrm{Hom}(\mathbf{y}(k), P) \cong P(k).$$

Thus either way we will have an iso $P \times [0 = 1] \cong [0 = 1]$ iff $P(k) \cong 1$.

The presheaf topos $\mathsf{Set}^{\mathbb{B}}$ is therefore the closed complement of the open subtopos

$$\mathsf{Set}^{\mathbb{B}_=}/_{[0=1]} \hookrightarrow \mathsf{Set}^{\mathbb{B}_=},$$

given by forcing the proposition $0 \neq 1$. Since $\mathsf{Set}^{\mathbb{B}_=}$ is clearly the classifying topos for *arbitrary* bipointed objects, say $\mathsf{Set}[B, 0, 1]$, the sheaf subtopos

$$\mathsf{Set}[B, 0 \neq 1] \simeq \mathsf{Set}^{\mathbb{B}} \hookrightarrow \mathsf{Set}^{\mathbb{B}_=} \simeq \mathsf{Set}[B, 0, 1]$$

classifies *strictly* bipointed objects, i.e. intervals, as claimed.   □

Corollary B.2 *The geometric realization functor to topological spaces*

$$R : \mathsf{cSet} \rightarrow \mathsf{Top}$$

preserves finite products, $R(X \times Y) \cong R(X) \times R(Y)$ and $R(1) \cong \{*\}$.

Proof Compose the inverse image of the classifying geometric morphism $\mathsf{sSets} \rightarrow \mathsf{cSet}$ of the 1-simplex Δ^1 with the standard geometric realization $\mathsf{sSets} \rightarrow \mathsf{Top}$, both of which preserve finite products.   □

Example B.3 (P. Aczel) The cubical set P of polynomials (say, over the integers), is defined by:

$$P_n = \{p(x_1, \ldots, x_n) \mid \text{polynomials in at most } x_1, \ldots, x_n\}$$

with the substitution map $s^* : P_n \rightarrow P_m$ taking $p(x_1, \ldots, x_n)$ to

$$s^* p(x_1, \ldots, x_n) = p\big(s(x_1), \ldots, s(x_n)\big),$$

for each bipointed map $s : [n] \rightarrow [m]$.

This cubical set P underlies a ring object in cSet, and the interval $\mathrm{I} = \mathsf{y}[1]$ embeds into it via the component maps

$$\eta_n : \mathrm{I}_n \to P_n$$

taking $v_i \in \Box([n], [1]) \cong \mathbb{B}\big([1], [n]\big) \cong \{0, x_1, \dots, x_n, 1\}$ to 0, 1, or the variable x_i, respectively, in P_n. The same is true for any algebraic theory $\mathbb{T}$ with two constants, such as Boolean algebras: there is a distinguished cubical $\mathbb{T}$-algebra $\mathcal{A}$, and a natural map $\eta : \mathrm{I} \to |\mathcal{A}|$ in cSet.

Indeed, let $\mathsf{cSet} = \mathsf{Set}[\mathcal{I}]$ as a classifying topos for intervals by Proposition B.1 with $\mathcal{I} = (1 \rightrightarrows \mathrm{I})$, and let

$$\mathsf{Set}[\mathbb{T}, \mathsf{flat}] = \mathsf{Set}^{\mathbb{T}^{\mathrm{op}}}$$

be the topos of presheaves on the Lawvere algebraic theory $\mathbb{T}$, which therefore classifies *flat* $\mathbb{T}$-algebras. There is a bipointed object $\mathcal{J} = (1 \rightrightarrows \mathrm{J})$ in $\mathbb{T}$, consisting of the generic $\mathbb{T}$-algebra and its two constants, which has a classifying functor $\mathrm{J}^\sharp : \Box \to \mathbb{T}$, inducing adjoint functors on presheaves,

$$\mathrm{J}_! \dashv \mathrm{J}^* \dashv \mathrm{J}_* : \mathsf{Set}[\mathcal{I}] = \mathsf{Set}^{\Box^{\mathrm{op}}} \longrightarrow \mathsf{Set}^{\mathbb{T}^{\mathrm{op}}} - \mathsf{Set}[\mathbb{T}, \mathsf{flat}] \,,$$

where $\mathrm{J}_! \circ \mathsf{y}_\Box = \mathsf{y}_{\mathbb{T}} \circ \mathrm{J}^\sharp$, with y the respective Yoneda embeddings.

We can then calculate,

$$\mathrm{J}^*\mathrm{J}_!(\mathrm{I})([n]) = \mathrm{J}^*\mathrm{J}_!(\mathsf{y}[1])([n])$$

$$= \mathrm{J}^*\mathsf{y}(\mathrm{J}^\sharp[1])([n])$$

$$= \mathsf{y}(\mathrm{J}^\sharp[1])(\mathrm{J}^\sharp[n])$$

$$= \mathbb{T}(\mathrm{J}^\sharp[n], \mathrm{J}^\sharp[1])$$

$$= \mathsf{Alg}_{\mathbb{T}}(\mathrm{J}^\sharp[1], \mathrm{J}^\sharp[n])$$

$$= \mathsf{Alg}_{\mathbb{T}}(F(1), F(n))$$

$$= |F(n)|,$$

where $|F(n)|$ is the underlying set of the free $\mathbb{T}$-algebra $F(n)$, the nth object of the Lawvere theory under its dual presentation $\mathbb{T}^{\mathrm{op}} \hookrightarrow \mathsf{Alg}_{\mathbb{T}}$. The unit of the $\mathrm{J}_! \dashv \mathrm{J}^*$ adjunction provides a natural map $\eta : \mathrm{I} \to \mathrm{J}^*\mathrm{J}_!(\mathrm{I})$, given pointwise by $\mathrm{I}_n \cong \{0, x_1, \dots, x_n, 1\} \longrightarrow |F(n)| \cong \mathrm{J}^*\mathrm{J}_!(\mathrm{I})_n$.

The cubical set of polynomials $P - \mathrm{J}^*\mathrm{J}_!(\mathrm{I})$ is thus indeed a cubical ring, with a map $\mathrm{I} \to P$, since $\mathrm{J}_!(\mathrm{I}) \cong \mathsf{y}_{\mathbb{T}} \mathrm{J}^\sharp([1]) \cong \mathsf{y}_{\mathbb{T}}(\mathrm{J})$ is a ring in $\mathsf{Set}[\mathbb{T}, \mathsf{flat}]$ and J^* is left exact. In fact, we learn thereby that P is flat.

Definition B.4 Let $\square \to \mathsf{Cat}$ be the unique product-preserving functor taking the interval $[1]$ to the one arrow category $2 = (0 \leq 1)$. This functor then takes $[n]$ to 2^n, the n-fold product in Cat, and maps $[m] \to [n]$ to the corresponding monotone functions $2^m \to 2^n$ of posets.[2] The *cubical nerve* functor

$$N : \mathsf{Cat} \to \mathsf{cSet}$$

is then defined by:

$$N(\mathbb{C})_n = \mathsf{Cat}(2^n, \mathbb{C}).$$

Thus $N(\mathbb{C})_0$ is the set of objects of $\mathbb{C}$; $N(\mathbb{C})_1$ is the set of arrows; $N(\mathbb{C})_2$ consists of all commutative squares; $N(\mathbb{C})_3$ all commutative cubes, etc.

Proposition B.5 *The cubical nerve N : $\mathsf{Cat} \to \mathsf{cSet}$ is full and faithful.*

Proof Given categories $\mathbb{C}$ and $\mathbb{D}$ and functors $F, G : \mathbb{C} \to \mathbb{D}$, suppose $F(f) \neq G(f)$ for some $f : A \to B$ in $\mathbb{C}$. Take $f^\sharp : 2 \to \mathbb{C}$ with image f. Then $N(F)_1(f^\sharp) = F(f) \neq G(f) = N(G)_1(f^\sharp)$, and so $N(F) \neq N(G) : N(\mathbb{C}) \to N(\mathbb{D})$. So N is faithful.

For fullness, let $\varphi : N(\mathbb{C}) \to N(\mathbb{D})$ be a natural transformation, and define a proposed functor $F : \mathbb{C} \to \mathbb{D}$ by

$$F_0 = \varphi_0 : \mathbb{C}_0 = N(\mathbb{C})_0 \to N(\mathbb{D})_0 = \mathbb{D}_0$$

$$F_1 = \varphi_1 : \mathbb{C}_1 = N(\mathbb{C})_1 \to N(\mathbb{D})_1 = \mathbb{D}_1.$$

We just need to show that F preserves identity arrows and composition. Consider the following diagram.

$$\begin{array}{ccc}
\mathsf{Cat}(2^1, \mathbb{C}) = N(\mathbb{C})_1 & \xrightarrow{\ F_1\ } & N(\mathbb{D})_1 = \mathsf{Cat}(2^1, \mathbb{D}) \\
{\scriptstyle !^*}\big\uparrow & & \big\uparrow {\scriptstyle !^*} \\
\mathsf{Cat}(2^0, \mathbb{C}) = N(\mathbb{C})_0 & \xrightarrow[\ F_0\ ]{} & N(\mathbb{D})_0 = \mathsf{Cat}(2^0, \mathbb{D}).
\end{array}$$

Here $!^* : \mathsf{Cat}(2^0, \mathbb{C}) \to \mathsf{Cat}(2, \mathbb{C})$ is precomposition with $! : 2 = 2^1 \to 2^0 = 1$, so the diagram commutes. But since $! : 2 \to 1$ is a functor,

$$\mathbb{C}_0 = \mathsf{Cat}(1, \mathbb{C}) \xrightarrow{\ !^*\ } \mathsf{Cat}(2, \mathbb{C}) = \mathbb{C}_1$$

---

[2] Thus factoring through the full subcategory $\square_{\wedge,\vee} \hookrightarrow \mathsf{Cat}$ of *Dedekind cubes*, mentioned above, which is the Lawvere algebraic theory of distributive lattices.

takes objects in $\mathbb{C}$ to their identity arrows. Thus F preserves identity arrows. Similarly, for composition, consider

$$\mathsf{Cat}(2^2,\mathbb{C}) = N(\mathbb{C})_2 \xrightarrow{\varphi_2} N(\mathbb{D})_2 = \mathsf{Cat}(2^2,\mathbb{D})$$
$$\downarrow{\scriptstyle d^*} \qquad\qquad\qquad\qquad \downarrow{\scriptstyle d^*}$$
$$\mathsf{Cat}(2,\mathbb{C}) = N(\mathbb{C})_1 \xrightarrow[F_1]{} N(\mathbb{D})_1 = \mathsf{Cat}(2,\mathbb{D}).$$

where $\varphi_2 : N(\mathbb{C})_2 \to N(\mathbb{D})_2$ is the action of φ on commutative squares of arrows, and $d^* : \mathsf{Cat}(2^2,\mathbb{C}) \to \mathsf{Cat}(2,\mathbb{C})$ is precomposition with the diagonal map $d : 2 \to 2^2 = 2 \times 2$, so the diagram commutes. For any composable $f : A \to B$ and $g : B \to C$ in $\mathbb{C}$ there is a commutative square

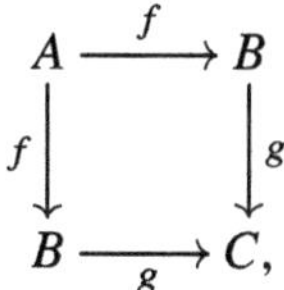

and the effect of $d^* : \mathsf{Cat}(2^2,\mathbb{C}) \to \mathsf{Cat}(2,\mathbb{C})$ on this square is exactly $g \circ f : A \to C$, and similarly for $d^* : \mathsf{Cat}(2^2,\mathbb{D}) \to \mathsf{Cat}(2,\mathbb{D})$. Thus the commutativity of the above diagram implies that F preserves composition. Since clearly $N(F) = \varphi$, we indeed have that N is also full. $\qquad\square$

References

1. Anel, M., Awodey, S., Barton, R.: A realizability ∞-topos (2026, in preparation)
2. Angiuli, C.: Computational semantics of cartesian cubical type theory. PhD thesis, Carnegie Mellon University (2019)
3. Angiuli, C., Harper, R., Wilson, T.: Computational higher-dimensional type theory. In: Proceedings of the 44th ACM SIGPLAN Symposium on Principles of Programming Languages, POPL 2017, pp. 680–693. ACM, New York (2017)
4. Angiuli, C., Hou (Favonia), K.B., Harper, R.: Cartesian cubical computational type theory: constructive reasoning with paths and equalities. In: Ghica, D., Jung, A. (eds.) 27th EACSL Annual Conference on Computer Science Logic (CSL 2018), Dagstuhl, vol. 119, pp. 6:1–6:17 (2018)
5. Angiuli, C., Brunerie, G., Coquand, T., Harper, R., (Favonia), K.B.H., Licata, D.R.: Syntax and models of cartesian cubical type theory. Math. Struct. Comput. Sci. **31**(4), 424–468 (2022)
6. Awodey, S.: A cubical model of homotopy type theory. Ann. Pure Appl. Logic **169**(12), 1270–1294 (2018)
7. Awodey, S.: Natural models of homotopy type theory. Math. Struct. Comput. Sci. **28**(2), 241–286 (2018)
8. Awodey, S.: Composition, filling, and fibrancy of the universe. Slides from a lecture at the meeting Foundations and Applications of Univalent Mathematics (2019)
9. Awodey, S.: A Quillen model structures on cubical sets. http://www.andrew.cmu.edu/user/awodey/talks/HoTT2019.pdf. Slides from a lecture at the conference Homotopy Type Theory 2019 (2019)
10. Awodey, S: On Hofmann–Streicher universes. Math. Struct. Comput. Sci. **34**(9), 894–910 (2024)
11. Awodey, S., Coquand, T.: Univalent foundations and the large-scale formalization of mathematics. The Institute Letter Summer 2013 (2013)
12. Awodey, S., Warren, M.A.: Homotopy theoretic models of identity types. Math. Proc. Camb. Philos. Soc. **146**, 45–55 (2009)
13. Awodey, S., Cavallo, E., Coquand, T., Riehl, E., Sattler, C.: The equivariant model structure on cubical sets. arXiv:2406.18497 (2024)
14. Awodey, S., Gambino, N., Hazratpour, S.: Kripke-Joyal forcing for type theory and uniform fibrations. Sel. Math. **30**(4), 74 (2024)
15. Barton, R.W.: A model 2-category of enriched combinatorial premodel categories. PhD thesis, Harvard University (2019)
16. Barton, R.:1 A short proof of the Frobenius property for generic fibrations. Math. Struct. Comput. Sci. **35**, e20 (2025)

S. Awodey, *Cartesian Cubical Model Categories*, Lecture Notes in Mathematics 2385, https://doi.org/10.1007/978-3-032-08730-0

17. Berger, C., Moerdijk, I.: On an extension of the notion of Reedy category. Math. Z. **269**, 1–28 (2008)
18. Bezem, M., Coquand, T.: A Kripke model for simplicial sets. Theor. Comput. Sci. **574**(C), 86–91 (2015)
19. Bezem, M., Coquand, T., Huber, S.: A model of type theory in cubical sets. In: 19th International Conference on Types for Proofs and Programs (TYPES 2013), vol. 26, pp. 107–128 (2014)
20. Bourke, J., Garner, R.: Algebraic weak factorisation systems I: accessible AWFS. J. Pure Appl. Algebra **220**, 108–147 (2016)
21. Brown, R.: Modelling and computing homotopy types: I. Indagationes Math. **29**(1), 459–482 (2018)
22. Brunerie, G: The James construction and $\pi_4(S^3)$. Institute for Advanced Study (2013). https://www.ias.edu/video/univalent/1213/0327-GuillaumeBrunerie
23. Brunerie, G., Licata, D.: A cubical infinite-dimensional type theory. Talk at Oxford Workshop on Homotopy Type Theory (2014)
24. Buchholtz, U., Morehouse, E.: Varieties of cubical sets. In: Höfner, P., Pous, D., Struth, G. (eds.) Relational and Algebraic Methods in Computer Science. Lecture Notes in Computer Science, vol. 10226, pp. 77–92. Springer, Berlin (2017)
25. Cavallo, E., Mörtberg, A., Swan, A.W.: Unifying cubical models of univalent type theory. In: Fernández, M., Muscholl, A. (eds.) 28th EACSL Annual Conference on Computer Science Logic (CSL 2020), Dagstuhl. Leibniz International Proceedings in Informatics (LIPIcs), vol. 152, pp. 14:1–14:17 (2020)
26. Cisinski, D.C.: Les préfaisceaux comme modèles des types d'homotopie. Astérisque **308**, 392 (2006)
27. Cisinski, D.C.: Higher Categories and Homotopical Algebra. Cambridge Studies in Advanced Mathematics, vol. 180. Cambridge University Press, Cambridge (2019)
28. Cohen, C., Coquand, T., Huber, S., Mörtberg, A.: Cubical type theory: a constructive interpretation of the univalence axiom. In: Uustalu, T. (ed.) 21st International Conference on Types for Proofs and Programs (TYPES 2015), Leibniz International Proceedings in Informatics, vol. 69, pp. 5:1–5:34 (2018)
29. Coquand, T.: Variations on cubical sets (diagonals version) (2014). Available from http://www.cse.chalmers.se/~coquand/diag.pdf
30. Coquand, T., Mannaa, B., Ruch, F.: Stack semantics of type theory. In: 2017 32nd Annual ACM/IEEE Symposium on Logic in Computer Science (LICS), pp. 1–11 (2017)
31. Dwyer, W.G., Kan, D.M.: Simplicial localizations of categories. J. Pure Appl. Algebra **17**(267–284) (1980)
32. Gambino, N., Garner, R.: The identity type weak factorisation system. Theor. Comput. Sci. **409**, 94–109 (2008)
33. Gambino, N., Henry, S.: Towards a constructive simplicial model of Univalent Foundations. J. Lond. Math. Soc. **105**, 1073–1109 (2022)
34. Gambino, N., Kock, J.: Polynomial functors and polynomial monads. Math. Proc. Camb. Philos. Soc. **154**(1), 153–192 (2013)
35. Gambino, N., Sattler, C.: The Frobenius condition, right properness, and uniform fibrations. J. Pure Appl. Algebra **221**(12), 3027–3068 (2017)
36. Garner, R.: Understanding the small object argument. Appl. Categ. Struct. **17**(3), 247–285 (2009)
37. Grandis, M., Mauri, L.: Cubical sets and their site. Theory Appl. Categ. **11**, 185–201 (2003)
38. Gratzer, D., Shulman, M., Sterling, J.: Strict universes for Grothendieck topoi. arXiv:220212012 (2022)
39. Grothendieck, A.: Pursuing stacks, unpublished (1983)

40. Hazratpour, S., Riehl, E.: A 2-categorical proof of Frobenius for fibrations defined from a generic point. Math. Struct. Comput. Sci. **34**(4), 258–280 (2024)
41. Hirschhorn, P.: Model Categories and Their Localizations. Mathematical Surveys and Monographs, No. 99. American Mathematical Society, Providence (2003)
42. Hofmann, M., Streicher, T.: Lifting Grothendieck universes, unpublished (1997)
43. Jardine, J.F.: Cubical homotopy theory: a beginning, unpublished (2002)
44. Johnstone, P.T.: Topos Theory. Academic Press, London (1977)
45. Joyal, A.: The Theory of Quasi-Categories and Its Applications. Quadern 45 vol. II. Centre de Recerca Matemàtica Barcelona (2008)
46. Joyal, A., Tierney, M.: An introduction to simplicial homotopy theory, unpublished (1999)
47. Joyal, A., Tierney, M.: Notes on Simplicial Homotopy Theory. CRM Publications, Montréal (2008)
48. Kan, D.M.: Abstract homotopy. I. Proc. Natl. Acad. Sci. USA **41**(12), 1092–1096 (1955)
49. Kan, D.M.: Abstract homotopy. II. Proc. Natl. Acad. Sci. USA **42**(5), 255–258 (1956)
50. Kapulkin, C., Lumsdaine, P.L.: The simplicial model of univalent foundations (after Voevodsky). J. Eur. Math. Soc. **23**, 2071–2126 (2021)
51. Kock, A.: Synthetic Differential Geometry, 2nd edn. London Mathematical Society Lecture Note Series. Cambridge University Press, Cambridge (2006)
52. Lawvere, W.: Left and right adjoint operations on spaces and data types. Theor. Comput. Sci. **316**, 105–111 (2004)
53. Licata, D.R., Orton, I., Pitts, A.M., Spitters, B.: Internal universes in models of homotopy type theory. In: Kirchner, H. (ed.) 3rd International Conference on Formal Structures for Computation and Deduction (FSCD 2018), Leibniz International Proceedings in Informatics (LIPIcs), vol. 108, pp. 22:1–22:17 (2018)
54. Ljungström, A.: The Brunerie number is −2. Homotopy Type Theory Blog (2022). https://homotopytypetheory.org/2022/06/09/the-brunerie-number-is-2
55. Lumsdaine, P.L.: Model structures from higher inductive types (2011). http://peterlefanulumsdaine.com/research/Lumsdaine-Model-strux-from-HITs.pdf
56. Lurie, J.: Higher Topos Theory. Annals of Mathematical Studies, vol. 170. Princeton University Press, Princeton (2009)
57. Maltsiniotis, G.: La catégorie cubique avec connexions est une catégorie test stricte. Homology Homotopy Appl. **11**(2), 309–326 (2009)
58. Marcelo Fiore, G.P., Turi, D.: Abstract syntax and variable binding. In: 14th Symposium on Logic in Computer Science (1999)
59. Moggi, E.: Notions of computation and monads. Inf. Comput. **93**(1), 55–92 (1991)
60. Orton, I., Pitts, A.M.: Axioms for Modelling Cubical Type Theory in a Topos. Logical Methods Comput. Sci. **14**(4), 1–33 (2018)
61. Parker, J.: Duality between cubes and bipointed sets. MS thesis in Logic, Computation and Methodology, Carnegie Mellon University (2015)
62. Quillen, D.G.: Homotopical Algebra. Lecture Notes in Mathematics, vol. 43. Springer, Berlin (1967)
63. Reedy, C.: Homotopy theory of model categories, unpublished (1974)
64. Riehl, E.: Algebraic model structures. N. Y. J. Math. **17**, 173–231 (2011)
65. Riehl, E.: Categorical Homotopy Theory. Cambridge University Press, Cambridge (2014)
66. Sattler, C: The equivalence extension property and model structures. ArXiv:1704.06911 (2017)
67. Shulman, M.: The univalence axiom for elegant reedy presheaves. Homology Homotopy Appl. **17**(2), 81–106 (2015)
68. Shulman, M.: All (∞, 1)-toposes have strict univalent universes. arXiv.1904.07004 (2019)
69. Swan, A.: W-types with reductions and the small object argument. arXiv.1802.07588 (2018)
70. Univalent Foundations Program, T.: Homotopy Type Theory: Univalent Foundations of Mathematics (2013). https://homotopytypetheory.org/book. Institute for Advanced Study
71. van den Berg, B., Faber, E.: Effective Kan Fibrations in Simplicial Sets. Lecture Notes in Mathematics, vol. 2321. Springer, Berlin (2022)

72. van den Berg, B., Garner, R.: Topological and simplicial models of identity types. ACM Trans. Comput. Logic **13**(1), 1–44 (2012)
73. Vezzosi, A., Mörtberg. A., Abel, A.: Cubical Agda: a dependently typed programming language with univalence and higher inductive types. Proc. ACM Programm. Lang. **3**(ICFP), 87:1–87:29 (2019)
74. Weber, M.: Yoneda structures from 2-toposes. Appl. Categ. Struct. **15**(3), 259–323 (2007)

Index

Editors in Chief: J.-M. Morel, B. Teissier;

Editorial Policy

1. Lecture Notes aim to report new developments in all areas of mathematics and their applications – quickly, informally and at a high level. Mathematical texts analysing new developments in modelling and numerical simulation are welcome.

   Manuscripts should be reasonably self-contained and rounded off. Thus they may, and often will, present not only results of the author but also related work by other people. They may be based on specialised lecture courses. Furthermore, the manuscripts should provide sufficient motivation, examples and applications. This clearly distinguishes Lecture Notes from journal articles or technical reports which normally are very concise. Articles intended for a journal but too long to be accepted by most journals, usually do not have this "lecture notes" character. For similar reasons it is unusual for doctoral theses to be accepted for the Lecture Notes series, though habilitation theses may be appropriate.

2. Besides monographs, multi-author manuscripts resulting from SUMMER SCHOOLS or similar INTENSIVE COURSES are welcome, provided their objective was held to present an active mathematical topic to an audience at the beginning or intermediate graduate level (a list of participants should be provided).

   The resulting manuscript should not be just a collection of course notes, but should require advance planning and coordination among the main lecturers. The subject matter should dictate the structure of the book. This structure should be motivated and explained in a scientific introduction, and the notation, references, index and formulation of results should be, if possible, unified by the editors. Each contribution should have an abstract and an introduction referring to the other contributions. In other words, more preparatory work must go into a multi-authored volume than simply assembling a disparate collection of papers, communicated at the event.

3. Manuscripts should be submitted either online at www.editorialmanager.com/lnm to Springer's mathematics editorial in Heidelberg, or electronically to one of the series editors. Authors should be aware that incomplete or insufficiently close-to-final manuscripts almost always result in longer refereeing times and nevertheless unclear referees' recommendations, making further refereeing of a final draft necessary. The strict minimum amount of material that will be considered should include a detailed outline describing the planned contents of each chapter, a bibliography and several sample chapters. Parallel submission of a manuscript to another publisher while under consideration for LNM is not acceptable and can lead to rejection.

4. In general, **monographs** will be sent out to at least 2 external referees for evaluation.

   A final decision to publish can be made only on the basis of the complete manuscript, however a refereeing process leading to a preliminary decision can be based on a pre-final or incomplete manuscript.

   Volume Editors of **multi-author works** are expected to arrange for the refereeing, to the usual scientific standards, of the individual contributions. If the resulting reports can be forwarded to the LNM Editorial Board, this is very helpful. If no reports are forwarded or if other questions remain unclear in respect of homogeneity etc, the series editors may wish to consult external referees for an overall evaluation of the volume.

5. Manuscripts should in general be submitted in English. Final manuscripts should contain at least 100 pages of mathematical text and should always include

   – a table of contents;
   – an informative introduction, with adequate motivation and perhaps some historical remarks: it should be accessible to a reader not intimately familiar with the topic treated;
   – a subject index: as a rule this is genuinely helpful for the reader.
   – For evaluation purposes, manuscripts should be submitted as pdf files.

6. Careful preparation of the manuscripts will help keep production time short besides ensuring satisfactory appearance of the finished book in print and online. After acceptance of the manuscript authors will be asked to prepare the final LaTeX source files (see LaTeX templates online: https://www.springer.com/gb/authors-editors/book-authors-editors/manuscriptpreparation/5636) plus the corresponding pdf- or zipped ps-file. The LaTeX source files are essential for producing the full-text online version of the book, see http://link.springer.com/bookseries/304 for the existing online volumes of LNM). The technical production of a Lecture Notes volume takes approximately 12 weeks. Additional instructions, if necessary, are available on request from lnm@springer.com.

7. Authors receive a total of 30 free copies of their volume and free access to their book on SpringerLink, but no royalties. They are entitled to a discount of 33.3 % on the price of Springer books purchased for their personal use, if ordering directly from Springer.

8. Commitment to publish is made by a *Publishing Agreement*; contributing authors of multiauthor books are requested to sign a *Consent to Publish form*. Springer-Verlag registers the copyright for each volume. Authors are free to reuse material contained in their LNM volumes in later publications: a brief written (or e-mail) request for formal permission is sufficient.

Addresses:
Professor Jean-Michel Morel, CMLA, École Normale Supérieure de Cachan, France
E-mail: moreljeanmichel@gmail.com

Professor Bernard Teissier, Equipe Géométrie et Dynamique,
Institut de Mathématiques de Jussieu – Paris Rive Gauche, Paris, France
E-mail: bernard.teissier@imj-prg.fr

Springer: Ute McCrory, Mathematics, Heidelberg, Germany,
E-mail: lnm@springer.com

MIX
Papier aus verantwortungsvollen Quellen
Paper from responsible sources
FSC® C105338

If you have any concerns about our products,
you can contact us on
ProductSafety@springernature.com

In case Publisher is established outside the EU,
the EU authorized representative is:
Springer Nature Customer Service Center GmbH
Europaplatz 3, 69115 Heidelberg, Germany

Printed by Libri Plureos GmbH
in Hamburg, Germany